ARTIFICIAL INTELLIGENCE AND NATIONAL DEFENSE

ARTIFICIAL INTELLIGENCE AND NATIONAL DEFENSE

•

Opportunity and Challenge

by

Paul E. Lehner, Ph.D.

TAB Professional and Reference Books

Division of TAB BOOKS Inc.
P.O. Box 40, Blue Ridge Summit, PA 17294

A Petrocelli book
FIRST EDITION
FIRST PRINTING

Library of Congress Cataloging in Publication Data

Lehner, Paul E.
Artificial intelligence and national defense : opportunity and challenge / by Paul E. Lehner.
p. cm.
Bibliography: p.
Includes index.
ISBN 0-8306-3235-2
1. Artificial intelligence—Military applications. 2. United States—Defenses. I. Title.
UG479.L45 1989
623'.028'563—dc19 89-4335
CIP

TAB BOOKS Inc. offers software for sale. For information and a catalog, please contact TAB Software Department, Blue Ridge Summit, PA 17294-0850.

Questions regarding the content of this book should be addressed to:

Reader Inquiry Branch
TAB BOOKS Inc.
Blue Ridge Summit, PA 17294-0214

Edited by Roman H. Gorski

The following material was reprinted with permission:
Figure 2-3 was reprinted from Woods, W.A. "Transition Network Grammars for Natural Language Analysis," *Communication of the ACM*, Vol. 13, 1970.
Figure 2-4 was reprinted from Rich, E. *Artificial Intelligence*, McGraw-Hill: New York, 1985.
Figure 4-5 was reprinted from Waltz, D. "Generating Semantic Descriptions from Drawings of Scenes with Shadows," in *The Psychology of Computer Vision*, ed. P. Winston, McGraw-Hill: New York, 1975.
Figure 8-6 was reprinted from Slagle, J.R. and Hamburger, H. "An Expert System for a Resource Allocation Problem," *Communication of the ACM*, Vol. 28, No. 9, 1985, pp. 994-1004.

To my wife and my love, Abby

Contents

Preface

Artificial intelligence (AI) is a discipline of computer science dedicated to the development of computational machines that exhibit intelligent behavior. During the past 20 years, the study of AI in machines has evolved from a purely academic activity to a major growth industry involved in both government and commercial applications. In particular, in the last five years, there has been an explosive growth in the number of AI systems oriented toward providing consumers with systems capable of giving practical advice on a variety of problems ranging from automechanics to corporate-level policy decision-making. A number of businesses have emerged that promote AI products as their primary enterprises, while many larger corporations have invested millions of dollars in the development of significant in-house AI capabilities.

Much of AI's success as a commercial enterprise can be attributed to the long-standing commitment of the U.S. Department of Defense

(DoD) and other government agencies to supporting the development of practical AI systems. Over the past decade, agencies such as the Defense Advanced Research Projects Agency (DARPA) have invested millions of dollars in the development of a practical AI technology that may contribute to the development of better systems for our national defense.

This book explores AI technology from the perspective of its potential contribution to our national defense, and investigates the extent to which this investment has paid off.

The exploration of AI as a potential contributor to our national defense capability is done from two perspectives: as an opportunity and as a challenge. The opportunity of AI is examined from the perspective of the variety of domains for which it appears that AI techniques are suited. As will be seen, they are quite numerous. The challenge of AI is that while there are many problems which AI systems could help solve, there still remain many significant impediments to implementing practical AI systems. Every effort is made to enumerate many of the difficulties developers of applied AI systems have faced and, for the most part, failed to overcome. The challenge to AI is to find an approach to overcoming these impediments.

This book is intended for two groups of readers. The first group includes those who do not have a background in AI, but have a need to understand how this technology might be applied, as well as its limitations. This group is composed primarily of potential consumers of AI technology who have or are about to make significant investments in the development of applied AI systems. For these readers, this book should provide a general understanding of what can reasonably be expected of AI as well as some of the more unrealistic expectations people often have.

The second group includes readers with some technical experience or training in AI who desire to further understand where this technology can be incorporated into our national defense capability. This group is comprised primarily of members of the AI growth industry who are looking for even more opportunities to sell their products.

This book is divided into two parts. Part I provides a layman's introduction to AI. The focus of this part is to provide readers, relatively new to AI, with a basic introduction to the various types of AI systems and their capabilities. Techniques in such areas as expert systems, natural language processing, computer vision, automated planning, intelligent computer-aided instruction, as well as several others, are presented. Part

II examines the applicability of AI to our national defense with a particular focus on command, control, communications and intelligence (C3I) applications, and as part of autonomous systems. The relationship of AI to the Strategic Defense Initiative (SDI) is briefly discussed. The applicability of AI is explored both from the perspective of prototype systems that have already been developed and from the perspective of the promises made about AI system capabilities in the near future. The reader should find it instructive to compare the limited capabilities of systems that have already been developed with the projected capabilities of some systems that have been promised within the next few years.

Acknowledgments

I owe a great debt of gratitude to those whose patience has allowed me to finish this book, especially to my wife, Abby, who always gave me encouragement when I needed it most.

I thank Steve Andriole for providing me with the opportunity to write this book, and my appreciation goes to PAR Government Systems Corporation for their support in this endeavor.

Finally, I thank two of my co-workers, Len Adelman and Paul Young, who often took it upon themselves to accept additional responsibilities so as to allow me more time to work on this manuscript.

Part 1
Artificial Intelligence Systems and Their Capabilities

Part I provides a general introduction to AI with relevance to our national defense, where "relevance" means that the techniques are presently being utilized in significant efforts to develop applied AI systems for our national defense. The goal of Part I is to provide readers without a significant background or experience in AI with a functional understanding of the basic capabilities and limitations of key AI technologies and systems. The details of these various classes of AI techniques are discussed only to the extent that they may provide such readers with a better understanding of functional capabilities and limitations.

1
Basics of Expert System Technology

Of the various types of AI systems, *expert systems* have received the most public exposure. Expert system technology is widely perceived as the AI technology with the most potential for the development of near-term applications. Literally hundreds of efforts are ongoing to attempt to exploit this potential.

Expert systems are computer-based consultants that provide advice to users in real-world complex problem domains. For example, one of the first expert systems to gain significant notoriety was MYCIN (Shortliffe, 1976). MYCIN provides expert advice to medical doctors on the diagnosis and treatment of various types of bacterial infections. It is considered an "expert" system because its procedures for diagnosing and recommending treatment are modeled after the procedures and decision-making heuristics employed by human experts (viz., experienced doctors). Emulating human expert behavior is often considered a defining characteristic of an expert system.

In this chapter, a basic introduction to expert system technology is provided. An extensive introduction can be found in Hayes-Roth, et al. (1983).

1.1 How Expert Systems Work

Expert systems are usually composed of two components: a *knowledge base* and an *inference engine*. The knowledge base is a data base or set of data bases that contain encoded problem-solving knowledge derived from human experts. The inference engine is a software package that functions to apply the encoded domain knowledge to problem-specific data to generate problem-specific recommendations. Figure 1-1 contrasts this approach with that of traditional programming where all procedures, algorithms, heuristics, rules, etc. needed for problem-solving are directly coded as software.

Frequently, expert knowledge is encoded in a knowledge base in the form of a set of IF-THEN declarations called *production rules*. These IF-THEN expressions are designed to provide a flexible approach to specifying situation-action pairs (that is, expressions that specify the action to be taken whenever the situation is believed true).

Figure 1-2 shows some examples of production rules from three well-known systems. These examples point out several features of the production rules that may be in a knowledge base. First, the complexity of these rules can range from a single IF-and-THEN clause to a complex combination of interrelated clauses defining the IF or THEN expressions. Second, the THEN expression can incorporate any of a variety of "actions," including estimates of the degree of belief that a hypothesis is true, suggesting to the user that he or she perform some action, or even automatically invoking a routine to perform some action (e.g., an automatic display command). Third, the THEN expressions of some rules may correspond to the IF expressions of other rules, thereby making it possible to create hierarchies of rules. These rule hierarchies, when represented graphically, are called *inference networks*. Figures 1-3 and 1-4 show several rules and their corresponding inference networks. Expert

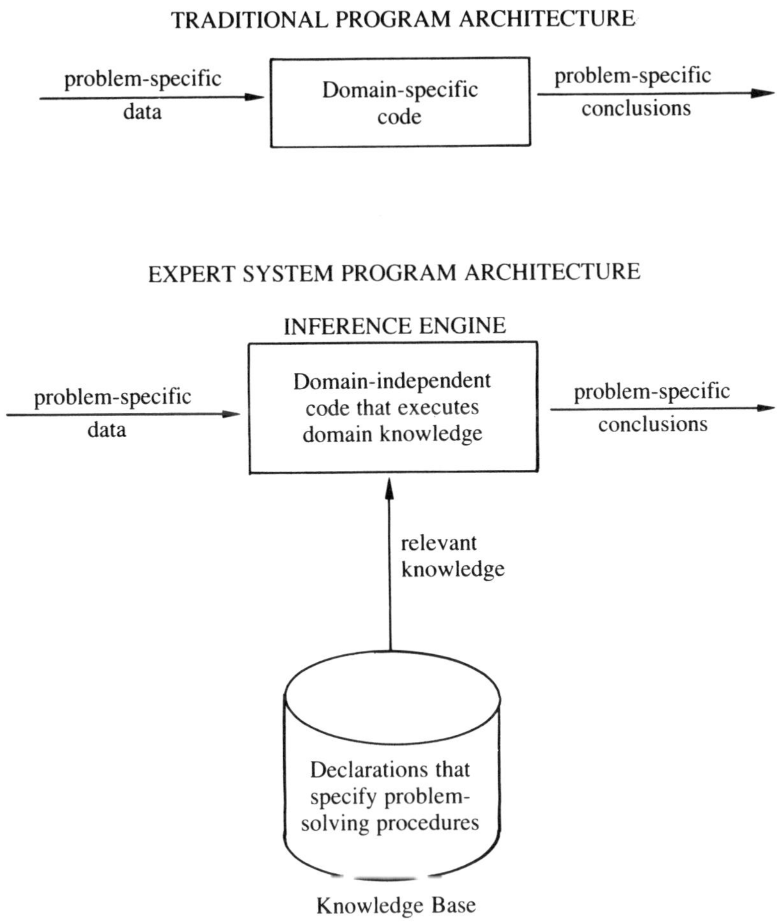

Figure 1-1. Traditional vs. expert system software architecture.

systems often consist of hundreds of rules combined into complex hierarchical networks.

The inference engine of an expert system performs a simple but very basic function. It applies the procedures implied by the rules in a knowledge base to specific problems. Consequently, the terms "rule applier" and "rule interpreter" have also been used to refer to inference engines. A simple inference engine might be one that performs the following steps:

SYSTEM: RI (configures VAX systems)

IF:
1) The current context is assigning devices to Unibus modules, and
2) There is an unassigned dural-port disk drive, and
3) The type of controller it requires is known, and
4) There are two such controllers, neither of which has any devices assigned to it, and
5) The number of devices that these controllers can support is known

THEN:
1) Assign the disk drive to each of the controllers, and
2) Note that the two controllers have been associated and that each supports one device.

SYSTEM: MYCIN (medical diagnosis)

IF:
1) The site of the culture is blood, and
2) The identity of the organism is not known with certainty, and
3) The stain of the organism is gramneg, and
4) The morphology of the organism is rod, and
5) The patient has been seriously burned

THEN: There is weakly suggestive evidence (0.4) that the identity of the organism is pseudomonas.

SYSTEM: PROSPECTOR (mineral exploration)

IF: There is hornblende pervasively altered to biotite

THEN: There is strong evidence (320, 0.001) for potassic zone alteration.

Figure 1-2. Sample production rules.

IF: The exhaust is smoky, and
The car is backfiring, and
There is a lack of power,

THEN: The carburetor fuel mix is too rich.

IF: There is a lack of power, and
There is a gray deposit on the spark plugs, and
The engine overheats,

THEN: The carburetor fuel mix is too weak.

IF: The carburetor fuel mix is too rich, or
The carburetor fuel mix is too weak,

THEN: The carburetor fuel mix needs to be adjusted.

Figure 1-3. Sample production rules.

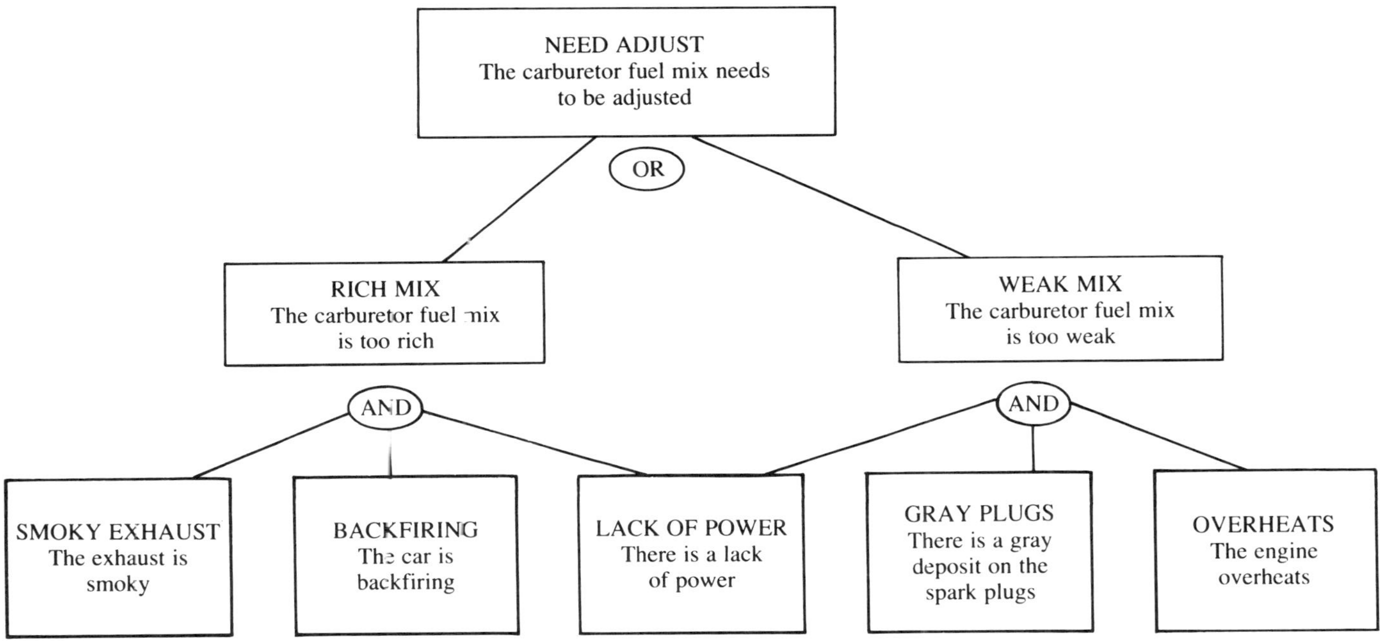

Figure 1-4. Sample inference network.

1. Search all clauses in the rule hierarchies to identify the *primitive* clauses, that is, IF clauses that do not correspond to THEN clauses in other rules.
2. For each primitive clause, ask the user if it is true or false.
3. For each THEN clause in the rule base, calculate if it is true or false and report the result to the user.

1.1.1 Alternative Control Strategies

The simple inference engine described above is theoretically sufficient for processing rule hierarchies of any size. However, as the number of rules in a rule base increases, the behavior of acquiring all available *evidence* or primitive values would become very inefficient, as well as frustrating to a user if he or she must enter all the data. To efficiently manage the application of domain knowledge to specific problems, inference engines apply a *control strategy* that carefully controls the order of rule activation.

One common strategy is to select a *goal hypothesis*, usually a top level hypothesis in the rule hierarchy, and to chain down the hierarchy one rule at a time to identify intermediate and primitive clauses that impact the selected goal hypothesis. The expert system then gathers data about evidence items specifically related to the goal hypothesis of interest. The medical diagnosis system MYCIN, for instance, considers only one type of bacterial infection at a time. It queries the user about relevant patient symptoms and test results until it acquires enough data to either accept or reject the diagnosis before it goes on to the next one. This approach of managing the utilization of rules is called a *goal driven* control strategy, inasmuch as it selectively pursues one goal at a time. It is also sometimes referred to as *backward chaining* or *consequent reasoning*.

A *data driven* control strategy is one in which rule activation is controlled by data available and not by the pursuit of particular goals. In the data driven approach the expert system awaits the input of new data. When new data is entered, the inference engine scans for rules that are impacted, applies the impacted rules to generate whatever conclusions it can, and then resumes waiting for input.

A data driven strategy is appropriate for situations where data becomes available intermittently over time, or for rule hierarchies in which

SYSTEM:	Currently trying to determine if the carburetor fuel mix needs to be adjusted.
SYSTEM:	Is the exhaust smoky?
USER:	No.
SYSTEM:	Is there a lack of power?
USER:	Yes.
SYSTEM:	Is there a gray deposit on the spark plugs?
USER:	Yes.
SYSTEM:	Does the engine overheat?
USER:	Yes.
SYSTEM:	Given the above evidence, conclusions are that the carburetor fuel mix is too weak. The carburetor fuel mix needs to be adjusted.

Figure 1-5. Sample consultation session with an expert system.

there are relatively few primitive clauses with a large number of goal hypotheses. The goal driven strategy is appropriate for situations where the number of goal hypotheses is small relative to the number of primitive clauses.

Figure 1-5 shows a simulated session with an expert system executing the rules in Figure 1-3. Note that this expert system uses strictly Boolean logic, and applied a goal driven strategy to select which rules to test.

1.1.2 Reasoning with Uncertainty

In most realistic domains, inference and decision-making occurs in the context of considerable uncertainty about the status of the problem being addressed. For an expert system, this implies that the status of the primitives will not be known with certainty. To deal with this problem, developers of expert systems generally incorporate some mechanism for incorporating degree-of-belief considerations into the rules. Rather than have rules that are strictly IF-THEN expressions, expert systems often include functions for calculating the "degree" to which the THEN conclusion is believed as a function of the degree to which the IF expression is believed. That is, rules of the form:

IF	A is believed with degree X
THEN	Conclude B with degree of belief f(X).

Various types of mathematics are used for calculating belief values. Bayesian techniques, based on probability theory, are the most popular. These techniques assign likely estimates to the various possible conclusions from which the probability that each conclusion is true can be derived. Mathematics from Fuzzy Set theory are sometimes used to calculate degrees of membership to a set or class. Evidential Reasoning techniques provide a mathematics that can be used to generate a non-probabilistic estimate of the degree of support for or against a conclusion. It is often the case that the equations in an expert system are not based on any theory at all, but rather reflect the ad hoc formulas that appear the most representative of how human experts assess a degree of belief in a conclusion.

1.1.3 Alternative Approaches to Representing Knowledge

In the above discussion of how expert systems operate, it was assumed that knowledge was encoded in a knowledge base in the form of production rules. This is not always the case. As AI has matured, other approaches to *knowledge representation* have been developed. Some of these alternative approaches are briefly summarized below.

Production Rules

As noted above, production rules are a series of IF-THEN expressions that specify the relationships between various situation-action pairs. Figure 1-6 shows a small rule base which will be used for comparison purposes.

IF: has feathers,
THEN: animal is a bird.

IF: flys and
lays eggs,
THEN: animal is a bird.

IF: animal is a bird, and
can't fly, and
color is black and white, and
can swim,
THEN: animal is a penguin.

Figure 1-6. Comparison rule base.

```
(OR  NOT (HAS_FEATHERS)
     ANIMAL_IS_BIRD)

(OR  (NOT (AND FLYS   LAYS_EGGS))
     ANIMAL_IS_BIRD)

(OR  (NOT (AND ANIMAL_IS_BIRD
               CAN'T_FLY
               COLOR_BLACK_AND_WHITE
               CAN   SWIM))
     ANIMAL_IS_PENGUIN)
```

Figure 1-7. Predicate logic equivalents of comparison rules.

Predicate Logic

Some expert systems have been built around the use of predicate logic formalisms for representing knowledge. In this approach, a set of predicates and the logical relationships between them are specified. These relationships correspond primarily to the typical AND, OR, IF-THEN, etc. expressions most people are familiar with. In this approach, the inference engine for applying knowledge may be much different than the production rule approach wherein an IF-THEN expression is processed by examining if the conditions under the IF part are true. If so, the THEN portion is processed. In the predicate logic approach, an [IF A, THEN B] expression translates into its logical equivalent [NOT(A) OR B], where the order relation (i.e., test A first) is lost. As a result, a predicate logic system could use the evidence "B is false" to conclude that "A is false," or it could use evidence that "A is true" to conclude that "B is true". Most expert systems built outside the U.S., particularly in Europe and Japan, use this approach. Figure 1-7 shows predicate logic equivalents to the rules in Figure 1-6.

Semantic Networks

Semantic networks are declarative representations that define concepts and objects in terms of relationships with other concepts and objects. Concepts and objects are described as nodes in a network where the arcs are labeled with types of relationships (e.g., a-kind-of, a-part-of). Semantic networks are sometimes embedded in systems that also have other knowledge representation techniques. The well-known geological analysis system PROSPECTOR, for instance, using both semantic networks and production rules. Figure 1-8 shows a semantic network representation of the rules in Figure 1-6.

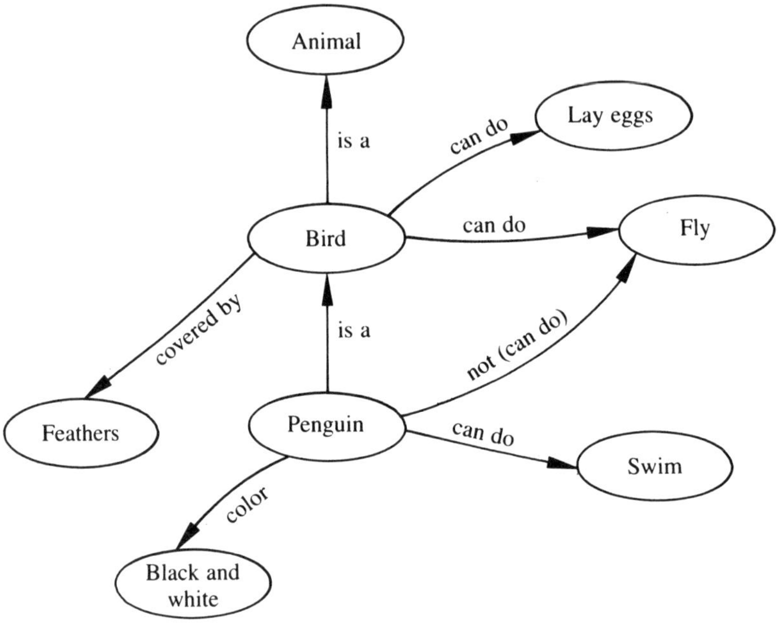

Figure 1-8. Semantic network equivalent to comparison rules.

Frames

Frames are static structures for describing various conceptual and physical objects. Within a frame there will be a set of slots that identify the dimensions by which an object can be characterized, and a set of attached procedures for filling those slots. Frames are usually organized by some higher order relationships that specify how various frames interrelate. Frame-based representations are becoming very common in AI-based decision support systems. The primary advantage is that they merge into a single framework both declarative knowledge, such as found in sematic nets, with procedural knowledge, such as found in production rules. Figure 1-9 shows a frame-based representation of the rules found in Figure 1-6.

Procedural Networks

Procedural networks define a partial order of actions over time. That is, they are networks where the nodes identify actions that may be taken, and the links are directed and specify the order in which action

BIRD
type of: Animal
covered by: Feathers
travels by: (Flies = default)
offspring by: Lays eggs

PENGUIN
type of: Bird
travels by: Swims
color: Black and white

Figure 1-9. Frame equivalent to comparison rules.

should occur. Procedural networks are frequently used to represent a plan of action (see Chapter 3). *Scripts* are template procedural networks that describe standard event sequences that frequently reoccur. Precedural networks and scripts are very common in expert systems that deal with mechanical, electrical, and electronic system diagnosis and repair. Figure 1-10 shows a sample procedural network.

For the most part, in developing an expert system, the selection of one approach to knowledge representation over another is more a matter of convenience than feasibility. As the examples in Figures 1-6 through 1-9 illustrate, more than one approach to knowledge representation can often be used to represent the same knowledge content. However, as will be discussed, the process of obtaining and encoding knowledge an be quite tedious and time-consuming. As a result, selection of a knowledge representation format that appears "natural" and appropriate for a domain can save a substantial amount of time over the duration of an expert system development project. Indeed, expert system developers often switch knowledge representation schemes midway through such projects. This can usually be done without too much difficulty.

1.1.4 Summary of Expert System Processes

There are a number of dimensions by which expert system software can vary. These variations are largely independent of knowledge content. That is, for a given knowledge content, any of a variety of expert

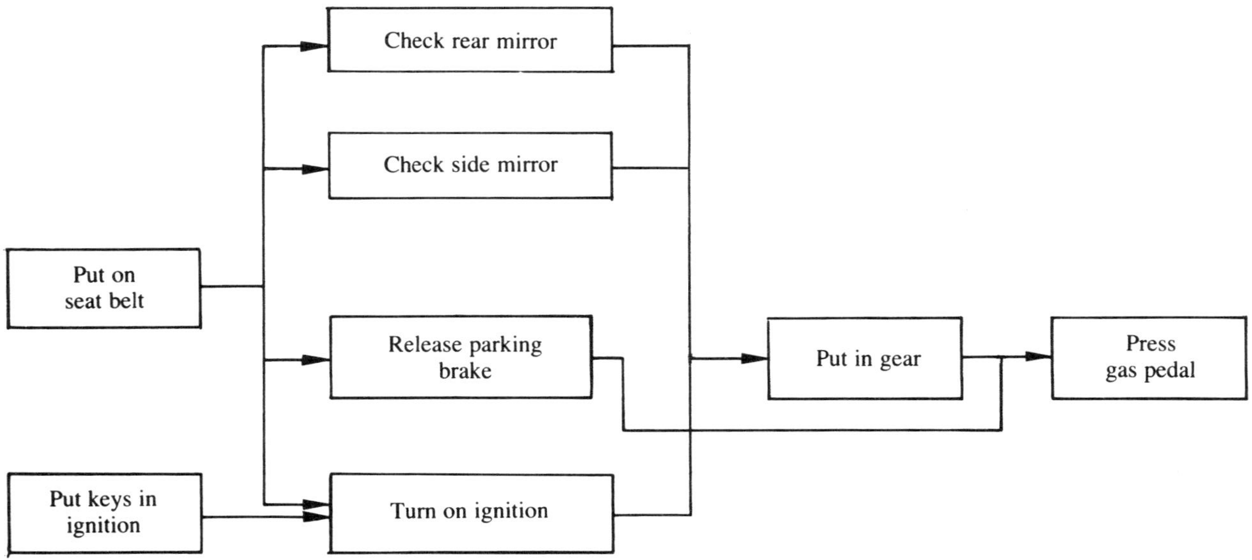

Figure 1-10. Procedural network for starting car.

system software packages can be developed to execute that content. Selection of a control strategy, equations for probabilistic reasoning, knowledge representation schemes, etc. may determine the *efficiency* of the execution of the knowledge, but often do not impact the *feasibility* of developing the expert system in the first place. The major feasibility issue revolves around the ability to define the knowledge content. Difficulties in the process of extracting and encoding knowledge and other general limitations of this technology are discussed in following sections.

1.2 The Problem of Knowledge Engineering

Expert systems can be described as computer-based consultants that emulate human expert reasoning in a problem domain. The process of extracting and encoding domain knowledge held by human expertise is called *knowledge engineering*. Given the state-of-the-science in AI today, knowledge engineering remains a time-consuming and labor-intensive process wherein an AI technologist, called a *knowledge engineer*, must repeatedly interview one or more human experts over a long time period (often years) to extract the heuristics to be encoded in the expert system knowledge base.

Knowledge engineering is essentially a process of instruction. The knowledge engineer must not only learn how the expert solves problems, but must learn it to such a level of detail that he or she can in turn provide precise instructions for a *tabulae rasa* machine. This is not an easy task inasmuch as human experts are often unaware of their own internal reasoning processes. Comments such as, "Once I understand the situation, the course of action is obvious," are common.

A variety of other factors contribute to a typically difficult knowledge engineering process. The following are examplary.

1.2.1 Significant Start-up Time

If a knowledge engineer is going to code a machine to exhibit expert problem-solving behavior, he/she must first achieve a minimal level of expertise with the domain before intelligently communicating with the domain expert. This usually requires significant background reading along with an initial set of interactive sessions with the domain expert.

1.2.2 Miscommunication Between Expert and Engineer

Since the engineer and expert are generally from very different backgrounds, the engineer and expert will often severely miscommunicate to the point where the engineer encodes a set of heuristics (with a corresponding organization of the knowledge base) that is on an entirely wrong track. Once the miscommunication is discovered, a major rework of the knowledge base is often required.

1.2.3 Limitations on Availability of Human Experts

People who achieve the classification of 'expert' are by definition very good at their job. They are usually much in demand and very busy. Interactive sessions with an expert must be planned long in advance and are generally infrequent. This problem compounds the previous one in as much as infrequent knowledge engineering sessions result in infrequent opportunities to correct miscommunications. This combination often produces very significant time delays.

Various estimates have been put forth for the amount of knowledge engineering it takes to build an expert system, with some ranging as high as two person-weeks per rule. While this situation is improving with experience, the process of encoding domain knowledge remains today much more of an art than a systematic science.

1.3 Limitations of Expert Systems Technology

Potentially, expert system technology provides a powerful set of tools for developing systems that can generate expert advice to users for solving important and complex problems. Unfortunately, while the potential is there, a number of practical difficulties exist that has resulted in relatively few successful applications of this technology. As has been noticed by many scientists, these difficulties significantly limit the types of problems that current expert system technology can successfully address. The following, drawn from Gevarter (1985), represent a typical characterization of these limitations.

- *Narrow domain of expertise.* Because building and maintaining large knowledge bases is difficult, only a few expert systems have been developed that cover a significant range of knowledge.
- *Limited knowledge-representation languages.* Certain kinds of knowledge, particularly those without immediate IF-THEN consequences, can be quite difficult to represent efficiently with current knowledge representation techniques.
- *Limited and stylized languages for input and output.* Users must describe problems in a strictly defined formal language. Explanations of outputs are limited and very stylized (e.g., rule traces).
- *Laborious construction.* At present, a knowledge engineer must extract knowledge from a human expert and laboriously build it into a knowledge base.
- *Need for a single expert to serve as "knowledge czar".* Expert system developers are currently limited to their ability to maintain consistency and resolve conflicts among overlapping items in the knowledge base. Thus, though several experts may contribute to a project, one expert must maintain absolute control to ensure the quality of the knowledge base.
- *Fragile behavior.* Most systems do not deal with problems at the boundaries of their knowledge. Furthermore, they generally do not have the necessary knowledge built into them to identify problems beyond their capacity. Consequently, even some of the best systems generate clearly false answers for problems outside their knowledge base. Even within their knowledge domain, they can be misled by unusual cases.
- *Knowledge engineer must operate the system.* Often, only the builders of an expert system or others with similar experience can effectively operate these AI systems. Questions to many of these systems must be carefully framed to extract reasonable solutions. Most system developers have not constructed friendly interfaces for their programs.

1.4 Tools for Building Expert Systems

A number of systems have been developed that are designed to support the process of expert system development. Most of these systems are problem-independent tools for expert system construction. The primary component of most of these tools is a generic inference engine that could, in principle, be applied to any of a variety of problem domains by entering domain-relevant knowledge into the knowledge base. Example systems of this type are listed in Table 1-1. The systems listed in the figure were all advertised in the Winter 1986 issue of *AI Magazine*. They represent only a small subset of a broad variety of software systems that provide expert system developers with an up-front rule application capability.

Despite the fact that a plethora of generalized inference engines exist,

Expert System Development Tool*	DEVELOPER OR DISTRIBUTOR
GURU™	Micro Data Base Systems, Inc.
Rule Master™	Radian Corporation
OPS5+	Artelligence, Inc.
KES II	Software Architecture and Engineering, Inc.
The Knowledge Workbench™	Silogic, Inc.
Personal Consultant™	Texas Instruments, Inc.
Exper Facts™	Exper Ielligence, Inc.
EXSYS	EXSYS, Inc.
ES/P ADVISOR	Expert Systems International Ltd.
KDS	KDS Corporation
OPS83	Production Systems Technologies, Inc.
DAISY	LITHP Systems
XSYS	California Intelligence
TOPSI	Dynamic Master Systems, Inc.
Expert Edge™	Human Edge
S.1 and M.1™	Teknowledge, Inc.

Table 1-1. Some Expert System Development Tools.

*All were advertised in the Winter 1986 issue of *AI Magazine*.

it is not at all clear (at least to this author) how much they contribute to the expert system development process. Even for a relatively small expert system, with say 200 to 300 rules, the most time-consuming aspect of development is the knowledge engineering process. The incremental cost of building a new inference engine that is tailored to the requirements of a problem domain is usually small relative to the knowledge engineering investment. Consequently, if a developer is oriented toward developing more than a small demonstration system, then budgeting the resources to develop a tailored engine may well be worth the investment.

On the other hand, if a small, proof-of-principle demonstration is desired, or if the system builder is very uncertain of the knowledge base requirements of the problem domain, then using an available general inference engine should significantly reduce the start-up cost of exploring the problem domain.

Another class of tools for supporting expert system construction provide direct assistance to the knowledge engineering process. These tools fall into three broad categories. The first category are tools that perform *internal rule checks*. For instance, a good rule base is noncircular in its reasoning (e.g., A→B, B→C, and C→A do not all occur), consistent in its conclusions (e.g., A→B and A→Not(B) are not both members), and nonredundant (e.g., A→C is not a member if A→B and B→C are members). Systems which check new rules against these criteria provide some feedback on the quality of the organization of a rule base. Several expert system development tools (e.g., EMYCIN) have feedback capabilities of this type.

A second category of knowledge engineering tools provide some *objective feedback on rule quality*. For instance, some systems maintain a library of test cases that permit a knowledge developer to assess the impact of new rules against these test cases. An example of this type of tool is discussed in Lemmer and Mitchell (1984; see Chapter 7 for summary), where the system they describe allows users to develop computer vision rule bases by testing proposed rule sets against a variety of images to determine if it is sufficient for detecting a searched-for object.

A third category of knowledge development tools perform *automatic rule generation*. Some systems, for instance, allow users to specify instances of a concept from which the system abstracts a set of rules for identifying future instances of that concept. The derived rules are usually much simpler in form than the complex rule hierarchies a knowledge engineer working with a domain expert may generate. The ACLS (Analog Concept Learning System) described in Chapter 6 provides some automatic rule generation capability.

1.5 Summary

Expert systems are widely perceived as the most practical and useful type of AI systems. As will be seen in Part II of this book, the expert systems are being developed that address problems in virtually every area of military decision-making. This is true despite the fact that many of the system development difficulties noted in Sections 1.2 and 1.3 often resurface.

Also, there is a key difference between the expert systems that are envisioned for our national defense and "traditional" systems such as MYCIN and PROSPECTOR that provide the foundation of expert system technology as it exists today. Whereas these traditional systems were all stand-alone systems, the expert systems envisioned for national defense are generally designed to be embedded in larger systems. In many cases, the role of the expert system is to be an *intelligent interface* between these larger systems and their users. At this point, it is premature to discuss the difference between stand-alone and intelligent interface expert systems and some of the unique technical problems associated with building intelligent interface expert systems. This discussion is found in Chapter 11 after examples of national defense applications of expert system technology are presented in Chapters 7 through 10.

2
Natural Language Processing

Within AI, natural language processing refers to machine understanding of a language that people use during the course of normal discourse. The field of research in machine-based natural language understanding involves several areas. A first-level decomposition that will be reflected here is the understanding between written text vs. the understanding of spoken speech. Written text understanding requires a machine with lexical, syntactic and semantic knowledge relevant to processing the language. An understanding of spoken speech requires this same knowledge, along with a perceptual processing capability sufficient to translate auditory sensory information into a form essentially isomorphic to written text.

A general introduction to AI techniques used to interpret individual sentences is in Section 2.1. Longer forms of discourse are examined in Sections 2.2 and 2.3. A more extensive introduction to national language processing can be found in Harris (1985). The chapter on natural language processing in Rich (1984) also provides a good introduction.

2.1 Understanding Individual Sentences

Processing single sentences requires two types of activities: interpreting the individual words in a sentence and integrating the individual word interpretations into a structure that represents the overall meaning of the sentence.

Interpreting individual words in a sentence requires that the machine, at a minimum, have access to a dictionary or *lexicon*. Unfortunately, most languages have many words with multiple possible meanings. A "diamond," for instance, can refer to a baseball field, a precious gem, or a geometrical shape. A bat can be for playing baseball or a type of flying mammal. The process of identifying a single interpretation for a word, when multiple interpretations are possible, is called *lexical disambiguation*. The disambiguation process is based on information from the rest of the sentence, where hopefully other words have been assigned precise meanings. ("Pick up your bat and play ball.") Lexical disambiguation is therefore the process of using understood parts of a sentence to generate a precise interpretation of other parts of a sentence. As a result, it is often embedded within the broader sentence interpretation process.

As stated above, sentence interpretation is a process that involves integrating the individual word interpretations into a structure that represents the overall meaning of the sentence. There are two general forms of this type of structural analysis: syntactic analysis and semantic analysis. *Syntactic analysis* refers to the transformation of a word sequence in a structure that characterizes how the words relate to each other. *Semantic analysis* is the process of assigning meaning to the structures and substructures that (usually) were created by the syntactic analysis. These two forms of analysis are discussed below.

2.1.1 Syntactic Analysis

Syntactic analysis involves a process called *parsing* for determining the underlying structure of a sentence. Sentence parsing requires a well-defined *grammar* that characterizes the set of possible structures for strings in the specified language. Once a grammar has been defined, a parser can assign a structure to any sentence that is grammatically correct. The resulting structure is often referred to as a *parse tree*. Figures 2-1 and 2-2 show a simple grammar for a subset of English and a parse tree based on this grammar. Note that each decomposition of a node in the parse tree is consistent with the grammatical rules shown in Figure 2-1.

S → NP VP
NP → ADJS N
NP → the NP
ADJS → ADJ
ADJS → ADJ ADJS
VP → V
VP → V NP
ADJ → old|new|small|big
N → bicycle|boy
V → broke|fixed

Figure 2-1. A simple grammar for a fragment of English.

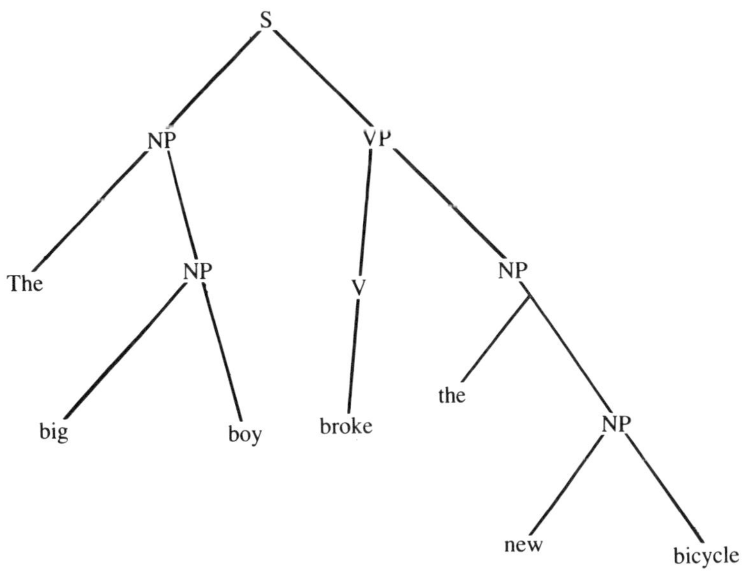

Figure 2-2. A sample parse tree.

Although the examples in the figures make parsing look straightforward, it can be a very complex process. Sentences are often structured such that early parts of the sentence can only be parsed after the rest of the sentence is parsed. Consider the sentence, "Have the students, who missed the exam, taken it today?" It is impossible to determine whether "Have" is an auxilliary or main verb until sentence parsing gets to "taken it today?".

To handle complex sentences such as this, natural language parsers generally employ some type of search mechanism to control development of the parse tree. Parsers differ in whether they use a top-down, bottom-up or combined approach to searching through the space of possible parse trees. In the top-down approach, the parser begins with a start symbol and applies the grammatical rules recursively until the parse tree decomposition results in terminal nodes that correspond to the specific components of the sentence being parsed. In the bottom-up approach, grammatical rules are applied backwards. This process starts by matching the postconditions of some of the grammatical rules to the sentence components and then recursively builds up higher levels of the tree until a single tree is found. A third approach is to combine bottom-up parsing with top-down filtering. In this approach, the parse tree is actually generated bottom-up, but as various nodes are added, top-down techniques are used to determine whether the evolving tree will ultimately be consistent with the grammar.

In addition to the search mechanism, a natural language parser requires a mechanism for handling ambiguities. For instance, the sentence that began with "Have the students . . ." has the embedded ambiguity of whether or not "Have" is a main or auxilliary verb. Techniques for parsing ambiguous words include the following:

- *Keep track of all paths*. Have the parser build a parse tree for all possible interpretations of the sentence as it is unfolding. This process continues until all but one parse tree has been rejected as invalid.
- *Follow best path and backtrack as needed*. Here the parser builds a single parse tree until it is complete, or until a word is found that is inconsistent with the tree as it is being built. Once an inconsistency is found, the parser backtracks to an earlier word, finds a new interpretation for that word, and continues as before.

- *Follow best path and patch tree as needed.* Like the previous approach, this approach builds a single tree until an inconsistency is found, but then resolves the inconsistency by modifying previous portions of the tree rather than backtracking.
- *Record ambiguities and resolve them later.* This is a "wait-and-see" approach that involves building partially specified trees. Early parts of the tree may have nodes that are not fully decomposed until parsing of latter parts of the sentence provides enough information to resolve the early ambiguity.

Most natural language systems will employ some combination of these techniques.

In addition to grammatical rules, natural languages also generally reflect a number of restrictions on the use of its grammar. That is, they generally reflect a restricted grammar. Exploiting these restrictions is of-

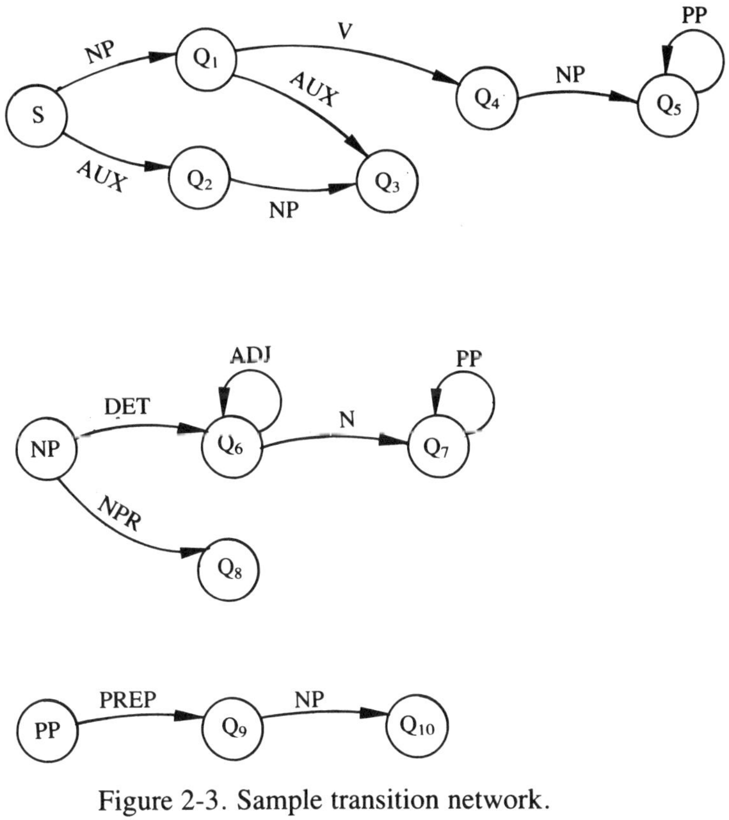

Figure 2-3. Sample transition network.

ten useful in limiting the amount of search required to parse a sentence. A common way of representing both the grammatical rules of a language and restrictions on the use of that grammar is with *transition networks*. Such networks identify for any element of a sentence the set of legal forms for the next element of the sentence, i.e., the set of legal transitions. Figure 2-3 provides a simple example of a transition network.

2.1.2 Semantic Analysis

The goal of semantic analysis is to determine the meaning of a sentence. Although it might at first appear that semantic processing can only occur after the syntactic structure of a sentence is uncovered, it is often the case that finding the correct syntactic interpretation depends on the semantics of the sentence. As a result, natural language processing systems often combine both syntactic and semantic analysis into a single common process. Two common approaches for combining syntactic analysis are (1) utilization of a semantic grammar and (2) the use of case grammars.

A *semantic grammar* is context-free wherein the elements and rules of the grammar reflect both syntactic and semantic relationships. Figure 2-4, for instance, shows a simplified semantic grammar drawn from the LADDER system described in Chapter 9. Note that the elements of this grammar incorporate semantic categories, rather than just syntactic categories like noun and verb phrases. Since a semantic grammar is defined in the same way as a purely syntactic grammar, syntactic parsing tech-

S → what is SHIP-PROPERTY of SHIP?
SHIP-PROPERTY → the SHIP-PROP | SHIP-PROP
SHIP-PROP → speed | length | draft | beam | type
SHIP → SHIP-NAME | the fastest SHIP2 |
    the biggest SHIP2 | SHIP2
SHIP-NAME → Kennedy | Kitty Hawk | Constellation | . . .
SHIP2 → COUNTRYS SHIP3 | SHIP3
SHIP3 → SHIPTYPE LOC | SHIPTYPE
SHIPTYPE → carrier | submarine | rowboat
COUNTRYS → American | French | British | Russian | . . .
LOC → in the Mediterranean | in the Pacific | . . .

Figure 2-4. Parts of a semantic grammar.

SENTENCE	CASE GRAMMAR INTERPRETATION
TOM broke the bicycle.	(BROKE (AGENT TOM) (OBJECT BICYCLE))
JOHN went to the store.	(WENT (AGENT TOM) (GOAL STORE))
Bob gave Mary a ring.	(GAVE (AGENT BOB) (BENEFICIARY MARY) (OBJECT RING))

Figure 2-5. Examples of case grammar interpretations.

niques can also be used to generate parse trees based on the semantic grammar. Once a parse tree is so generated, it will show the relationships among the semantic elements of a sentence and thereby provide a complete interpretation of that sentence.

One disadvantage of the semantic grammar approach is that the number of rules in the grammar may be prohibitively large for even very limited domains of discourse. This is because all acceptable relationships among various semantic elements for the domain of discourse must be represented in the grammar. Furthermore, even when an extensive semantic grammar is generated, its size will result in a lengthy and expensive parsing process.

Case grammars provide an approach to sentence interpretation that translates the results of a syntactic parse into a deeper, more semantically-oriented structure. Figure 2-5 exhibits several examples of case grammar interpretations. Note that these interpretations all revolve around verbs and the relationship of those verbs to their arguments. Listed in Figure 2-6 are some of the semantic categories (i.e., deep cases) that are found in all case grammars.

Agent	—	instigator of action
Dative	—	entity affected by action
Instrument	—	object used to cause event
Factitive	—	result of the event
Location	—	place of event
Source	—	starting location of a movement
Goal	—	ending location of a movement
Beneficiary	—	the benefactor of the event
Time	—	time of event occurrence
Object	—	entity changed on/acted upon

Figure 2-6. Some common elements of case grammars.

Sentence parsing with a case grammar operates in a manner similar to the syntactic parsers discussed previously. However, since case grammars have a very detailed and explicit representation of how verbs may relate to their arguments, this knowledge is used to guide the parsing process. In particular, once the verb of a sentence is isolated, it is used to predict which noun phrases will also occur in the sentence, and then the relationship of each phrase to the rest of the sentence. This added knowledge about semantically meaningful relationships limits the syntactic parsing process to only exploring syntactic structures that are consistent with the case grammar.

A third approach to semantic analysis is through the use of the *conceptual dependency* representation. This approach is similar to that of a case grammar in that it provides an underlying semantic representation that is independent of the structure of the original sentence. However, the conceptual dependency representation is a much deeper and lower level representation than that of a case grammar. This deeper level is achieved by defining a set of very primitive actions that, when taken in various combinations, correspond to the intended meaning of most verbs. For instance, the word "gave" would require the following conceptual dependency elements to be included:

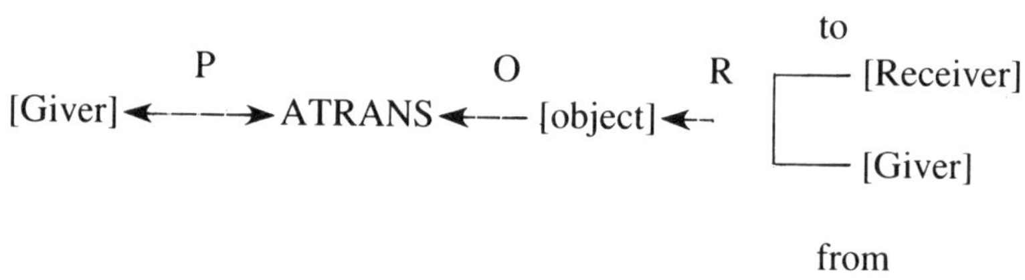

The symbols in this example have the following meanings:

- Arrows indicate the direction of the dependency
- Double arrows indicate a two-way link between actor and action
- P indicates past tense
- ATRANS is one of the primitive acts used by the theory to indicate transfer of possession
- O indicates the object case relation
- R indicates the recipient case relation

Figures 2-7 and 2-8 provide a brief description of some elements of the conceptual dependency representation and the use of that representation for the deep semantic interpretation of some sentences.

2.1.3 Pragmatic Analysis

In addition to a strictly syntactic and semantic analysis, sentence interpretation often requires some form of pragmatic analysis to determine

PTRANS	Transfer of location of an object
ATRANS	Transfer of a relationship (e.g., give)
PROPEL	Apply physical force to an object
MOVE	Self-movement of a body part (e.g., throw)
MTRANS	Transfer of information (e.g., tell)
MBUILD	Build new information (e.g., conclude)
SPEAK	Produce sounds
ATTEND	Focus sense organ (e.g., watch)
INGEST	Ingest an object into body

Figure 2-7. Some primitive actions in conceptual dependency.

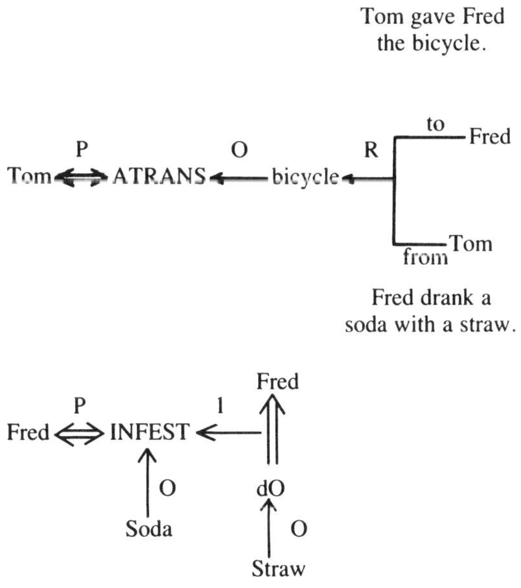

Figure 2-8. Two examples of conceptual dependency representation.

why a sentence was generated. For instance, "Do you know what time it is?" is clearly a request for the time. A strict syntactic/semantic analysis approach would not normally reveal such an implicit request. Understanding the pragmatics of a sentence usually requires understanding the broader context that a sentence is within, either as part of a collection of sentences or as part of an interactive dialogue. These topics are covered in the following two sections.

2.2 Understanding Multiple Sentences

Sentence understanding involves generating a structure representation that shows the relationships among various parts of the sentence. The same is true of understanding a collection of sentences. For instance, paragraph understanding can also be done by generating a structure that represents the paragraph elements and their interrelationships. Three types of structures commonly used for understanding written text are semantic networks, goal structures, and scripts/schemas.

Semantic networks characterize relationships between words. Such networks provide a convenient mechanism for representing the contents of declarative sentences. In addition, semantic network representations of the relationship between objects in a domain of discourse can effectively assist sentence understanding. Once a sentence has brought into "focus" specific elements of a domain of discourse, then future references to those elements can be understood quickly (see Figure 2-9).

While semantic networks provide a good background representation for understanding texts dealing with events and physical objects, they do not provide a sufficient structure for understanding stories that deal with human actions and motivations. Interpreting this type of text requires an understanding of the goals of the actors and their plans for achieving those goals. Consider the following story:

> Fred was hungry. He asked a man where the nearest restaurant was.

To understand this story, one needs to recognize that the primary actor had a goal of obtaining food, and a plan for achieving that goal, viz., go to a restaurant.

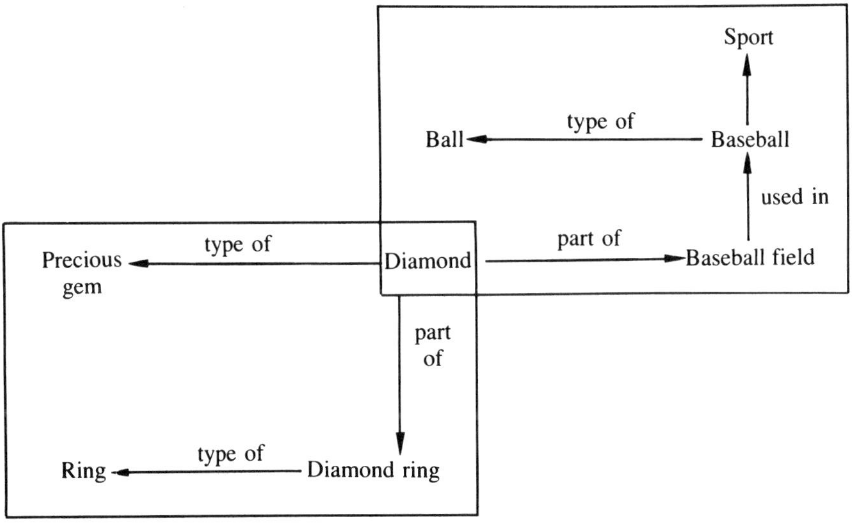

Figure 2-9. Example of semantic network focusing. The interpretation of "Diamond" depends on the space in focus.

One approach to representing goal structures is found in the story understanding system PAM (Planner Applier Mechanism). PAM incorporates a knowledge base of goals and goal-seeking actions represented as conceptual dependencies. It performs story understanding by first passing a text to the conceptual dependency parser, and then using the resulting representation to identify goals and goal-seeking actions embedded in the story. When all the actions in a story can be explained by a consistent set of goals and goal-seeking actions, then the story is considered to be understood (see Wilensky, 1988).

A third approach to text understanding is based on the observation that many event sequences frequently reoccur. For instance, the basic sequence of events that occurs when one goes to a restaurant is static. A *script* represents a pattern of events that will typically occur in certain contexts. Scripts are frame-based structures that contain slots that identify the elements of a stereotypical event sequence (Schank and Abelson, 1977). Once a script has been identified as appropriate for a segment of text, then it can be used to understand meanings in the text that were not explicitly stated. Consider the following story:

John went out to lunch. He sat down at a table and called the waitress. The waitress brought him a menu, and he ordered a hamburger. Afterwards, John went back to work.

While to a human reader this story is perfectly simple, to a machine there are many gaps. The following are some questions that might be asked. Why did the waitress bring John a menu? Did John eat anything? What did he eat? Did he pay for lunch? Questions like these cannot be understood without a fuller understanding of what normally occurs when one goes to a restaurant. Applying this general knowledge to a specific story is done by (1) identifying a relevant script, and (2) filling in the script slots with the specifics of the story. Slots defining events that are not filled-in are still assumed to have occurred, just the specifics remain in doubt. For the example given above, mapping the text segment to the restaurant script can then be used to more fully understand the text and answer the above questions.

Note that having a script that describes the normal course of events in certain contexts can also be used to detect and understand abnormal occurrences. Consider the following story:

John went to a restaurant. He was shown to his table. He ordered a large steak. He sat there and waited a long time. He got mad and left.

To understand why John got mad and left, it is necessary to understand that the normal event sequence is to order a meal and then be served that meal. Consequently, a script-based test understanding system, when given a text segment such as this, would be able to correctly answer the question, "Why did John get mad and leave the restaurant?"

2.3 Carrying on a Dialogue

Given that we have a machine that can understand natural language text, the next step would appear to be a machine that can carry on an active natural language dialogue. As with everything else, dialogue understanding poses some interesting and unique difficulties. These include the use of indirect speech acts where people say less than they really mean, implicit presuppositions where people ask questions that presuppose incorrect facts, anaphoric references that require interpretations based on an earlier portion of the dialogue, and the frequent use of fragments that do not represent grammatically correct sentences. These difficulties are further compounded by the fact that most dialogues represent a mixed initiative (Carbonell, 1970), where the topic and focus of conversation frequently shifts, and the control of these shifts can occur by either party.

Dialogue understanding systems use one or more of the following techniques to handle these difficulties:

- *Dialogue goal understanding*. Each utterance generated by the user must be interpreted in terms of the purpose of the act (answering a question, requesting information, issuing a system command, clarifying a previous utterance, etc.). A dialogue understanding system must be able to recognize implicit goals and respond accordingly.
- *Maintain current focus of dialogue*. Dialogues involve frequent changes of topic with little warning. Dialogue understanding systems must be able to maintain and represent the current focus of conversation. One technique for doing this is the use of focus spaces in semantic networks. These spaces isolate subsets of a network that represent the present domain of the dialogue and the set of objects and events relevant to the present focus.
- *Model each participant's beliefs*. Frequently, dialogue participants differ on underlying assumptions and beliefs relevant to a conversation. Indeed, many dialogues can be characterized as attempts to reconcile different beliefs. To handle this aspect of interactive dialogues, machines must have some model of both the machine's and human's dialogue-relevant beliefs. Semantic networks, for instance, can be used to represent the space of all dialogue relevant beliefs, while focus spaces can be used to represent the subspace of machine and user beliefs.
- *Understand shared rules of conversation*. In a dialogue with two cooperating parties, certain rules of conversation may be assumed to be valid. For instance, if A makes a request of B, then it is a sincere request. That is, (1) A wants B to do the request,

(2) A believes B can reasonably do the request, (3) A believes B would be willing to do the request, and (4) A believes B would not have done the requested action anyway. A great deal of implicit information can be drawn from the common rule that requests will be sincere.

2.4 Summary

The value of a natural language processing system is that it provides the opportunity for user/computer dialogues that are in terms and references familiar to the user. Presumably, this dialogue is about something of interest to both the computer and human, which in most applications implies that natural language processing systems provide some type of interface between the user and a computer-based system or data base computer. The development of natural language interfaces is consequently the primary application of this technology to date. Examples of this type of application in the national defense arena will be discussed in Part 2.

3
Automated Planning Systems*

In AI, a plan is defined as a predefined sequence of actions. *Automated planning* refers to the automated ability to define a sequence of actions that will achieve a specific goal. In the past 30 years, much of the research in AI has been dedicated to the development of systems for generating plans of action for various kinds of agents faced with numerous, complex, and conflicting goals.

In this chapter, two fields of research in AI are overviewed: action planning and game playing. *Action planning* deals with the automated planning of action sequences for such problem classes as robot maneuvers, task sequencing and others. The second field, *computer game playing*, includes systems for playing games such as chess, Go, checkers, Othello, etc. Combining these two research areas as a single topic is

*This chapter is adapted from Young and Lehner (1986).

somewhat unusual, but the contention can be made that *military planning is inherently adversarial* in nature, and consequently can only be addressed by merging both technologies. The discussion of computer game playing is limited to those topics that have relevance to military planning problems. The last section of this chapter overviews a theory of combining AI planning and game playing techniques into a single *adversarial planning* system.

3.1 Action Planning

There is an immense literature on the subject of AI techniques for action planning. A good summary of the different approaches can be found in Cohen and Feigenbaum (1982), while Christiansen (1985) has developed a general bibliography which includes the principle contributions to this field. Here we will present a broad outline of concepts and techniques.

3.1.1 Action Planning as Goal Tree Generation

Within action planning research, a common experimental planning task involves a robot (real or hypothetical) that exists in some type of world consisting of blocks. The robot is given tasks which require it to move the blocks into a particular configuration. Before starting, the robot is required to "figure out" what actions need to be taken—that is, to make a plan.

Some of the early planners (e.g., Fikes, et al. 1972) would solve problems such as this through a *backward chaining* procedure. In this approach, the planner is given a set of goals that are sufficient to describe the goal state, a description of the initial state of the world, and a set of permissible operators (actions) that can modify the world state. The planner then looks at each goal, matches it against the present world state and, if it is not already true, looks for an operator that can achieve it. If the operator cannot be applied because a precondition of its use is not satisfied, then the unsatisfied precondition becomes a new goal to be achieved. This process is recursively repeated until all goals are satisfied.

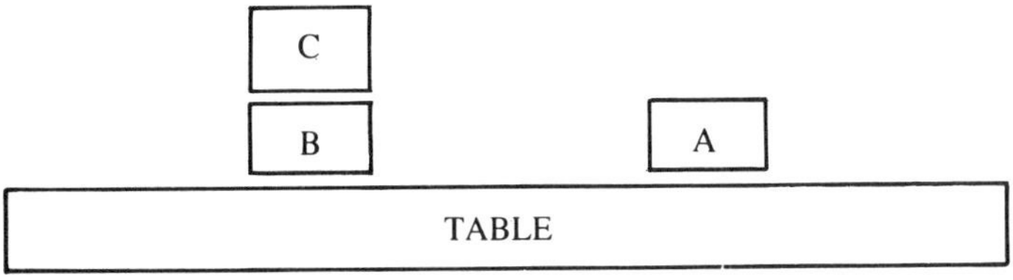

Figure 3-1. Initial configuration for blocks world problem.

To illustrate this approach, assume that we have a robot that exists in a world consisting of three square blocks and a table. Assume further that the blocks are initially configured as in Figure 3-1. The robot is capable of only one action, PUTON(X,Y), which means pick up block X and put it on top of object Y. Assume even further that the robot has been given the task of moving the blocks so that block A is on top of B which is on top of C (see Figure 3-2).

This goal state can be described by the conjunctions of goals ON (B,C) and ON(A,B). Assume that our robot has been given these goals, and it begins by trying to attain the first one, ON(B,C). To obtain this goal, the action PUTON(B,C) must be executed. However, this cannot be executed until the precondition that "C is clear" is true. Consequently, CLEAR(C) becomes a new goal to be achieved. PUTON (C, Table) can directly achieve this and is consequently the first action in the plan. Once C is clear, PUTON(B,C) is executable and therefore becomes the second action. At this point, ON(B,C) has been achieved, and ON(A,B) which is executable. Consequently, this becomes the third action in the plan, at which point the plan is complete.

One way to record this planning process is by the use of a *hierarchical goal tree* where each goal, subgoal and action that is part of the plan is recorded as a separate node. For instance, the tree in Figure 3-3 corre-

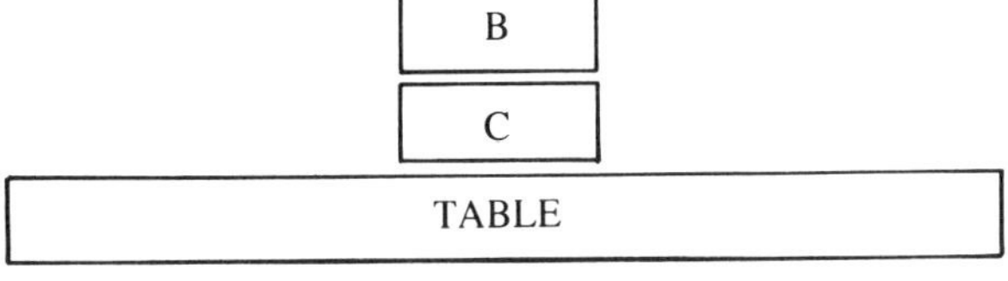

Figure 3-2. Final configuration for blocks world problem.

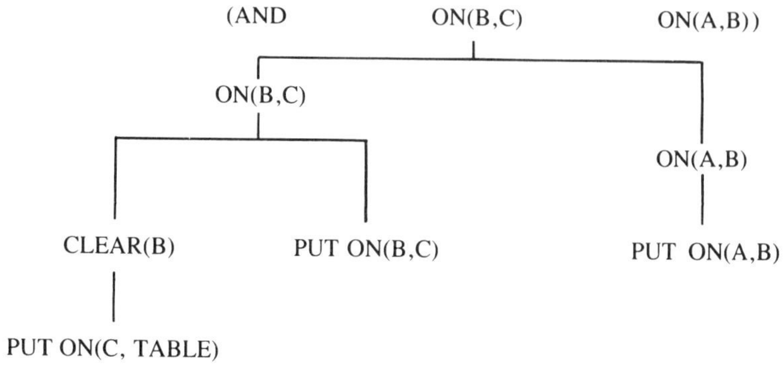

Figure 3-3. Plan represented as a goal tree.

sponds to the plan generated above. There are two features of this goal tree that bear notice. First, the sequence of actions in the final plan are contained in the terminal nodes. They are simply read from left to right. Second, by examining the ancestors for each action, we can determine why each action needs to be taken. For instance, the action PUTON(C,Table) is taken to CLEAR(C), which is necessary before ON(B,C) can be achieved, which is one of the original goals to be achieved. An action plan represented as a goal tree, therefore, not only consists of the actions that need to be executed, but also of an explanation of why each action is needed. It is also important to note that this goal tree was generated through a recursive *depth-first* expansion of the tree. For instance, before adding ON(A,B) to the tree, all descendents of ON(B,C) were added.

3.1.2 Hierarchical Planning

In general, most AI planning systems employ a hierarchical model for both plan representation and generation. Planners that employ *hierarchical plan generation* techniques maintain several versions of their plan for achieving a goal, each version being at a different level of abstraction or detail. The advantage of planning by this approach is that one is able to avoid dealing with the more detailed aspects of the plan until the general framework is completed. As an example, consider the case of a planner who wants to fly from Washington to Los Angeles. There is little value in determining the details of getting to an airport (which street to take, which time to leave, etc.) until the more abstract matter of determining which flight to take, and therefore which airport to leave from, has been worked out. In our previous blocks world example, a hierarchical planner would have added both ON(B,C) and ON(A,B) to the goal tree before determining the actions needed to achieve each of these individually.

3.1.3 Procedural Networks

One popular generalization of the hierarchical goal tree structure is the *procedural network* (Sacerdoti, 1975). Procedural networks only define a *partial ordering* of actions in a plan, and leave unordered action that can be executed in any order without affecting the plan outcome. To see how procedural nets can be used to represent plans of action, consider the planning problem depicted in Figure 3-4. Here, the problem is to capture the city with the given units which are initially on the

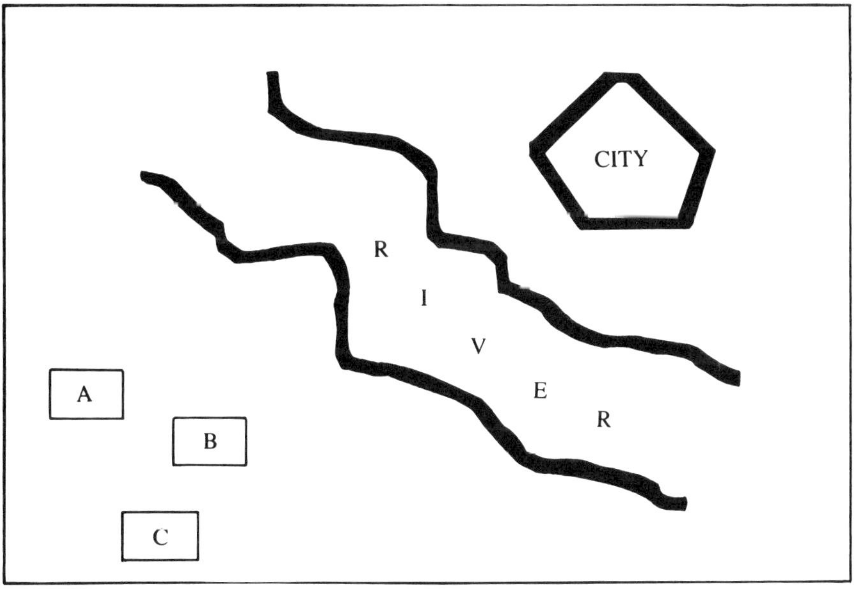

Figure 3-4. Sample planning problem—capture city using units A, B and C.

wrong side of the river. The objective can be expressed in the form of a single node plan as shown in Figure 3-5. This single node can be expanded into a more detailed plan by breaking the conjunction into its components; the resulting decomposition is in Figure 3-6. The three subproblems are shown in parallel to indicate that there is no a priori commitment to a temporal sequence. Obviously, however, for the goal to be achieved, the subproblems must be solved in a particular order. In this case, the city cannot be attacked until the units are across the river and the units cannot get across the river until the bridge is built. Such an interaction between subproblems has been termed a *conflict* by Sacerdoti. Conflicts are resolved by the application of a device known as a *critic* which looks for special kinds of interactions in a developing plan and adjusts the plan accordingly. Critics, because they are intended to be applicable to multiple planning situations, are necessarily predefined.

The conflicts in the sample plan developed so far would be resolved by a critic (or critics) that knew that a precondition for attacking a city is being near the city, and a precondition for being near something which is on the other side of a river would be the use/construction of a bridge to get across. After resolution of the conflicts, the CAPTURE_CITY procedural network would look as shown in Figure 3-7.

LEVEL ONE

(ACHIEVE (AND (ATTACK CITY) (CROSS RIVER) (BUILD BRIDGE)))

Figure 3-5. First decomposition level of planning problem.

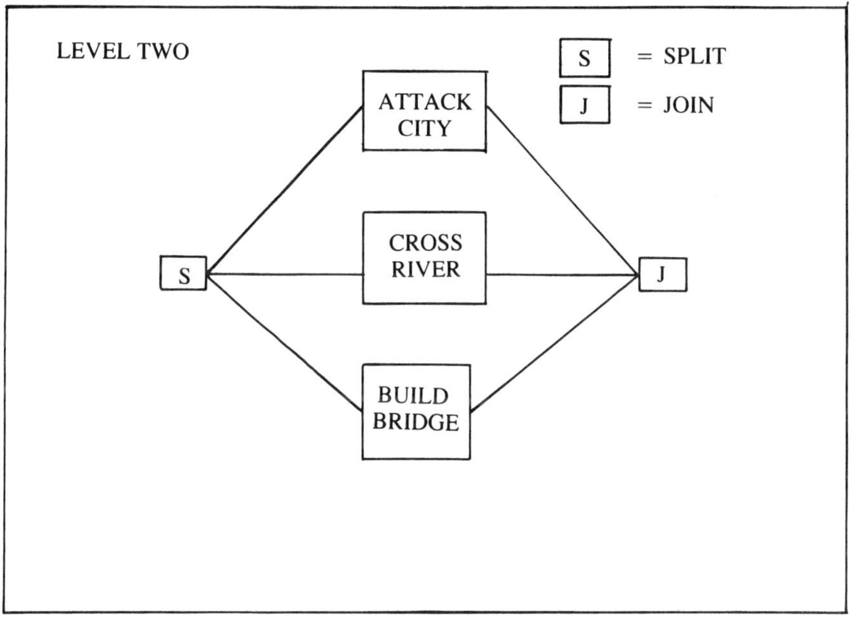

Figure 3-6. Second decomposition level of planning problem.

A close look at the initial problem (Figure 3-4) and the current plan (Figure 3-7) shows that there is still some ambiguity in that there are three distinct units to be moved across the river. Therefore, the plan can be further decomposed into a new set of subproblems, each representing a step for moving each of the three units across the river. The new, more complex plan is shown in Figure 3-8.

Once again, the subproblems are shown in parallel, and as long as the three units represent the same or similar resources, this plan is fine. In a real military planning situation, however, unit A may be artillery, unit B infantry, and unit C may be armor. In this case, the order of sending the units across may make a great deal of difference to the success of the operation. If the infantry (unit B) crosses first, they will be subjected to the full fury of the enemy's resistance while the armor (unit C) will be following uselessly behind. We have thus discovered a new conflict among subproblems in our plan; this conflict will have to be resolved by yet another critic. One possible criticism might be represented by a rule that armor always preceded infantry and that the artillery followed the infantry. Another possibility might be a critic which knew that within a

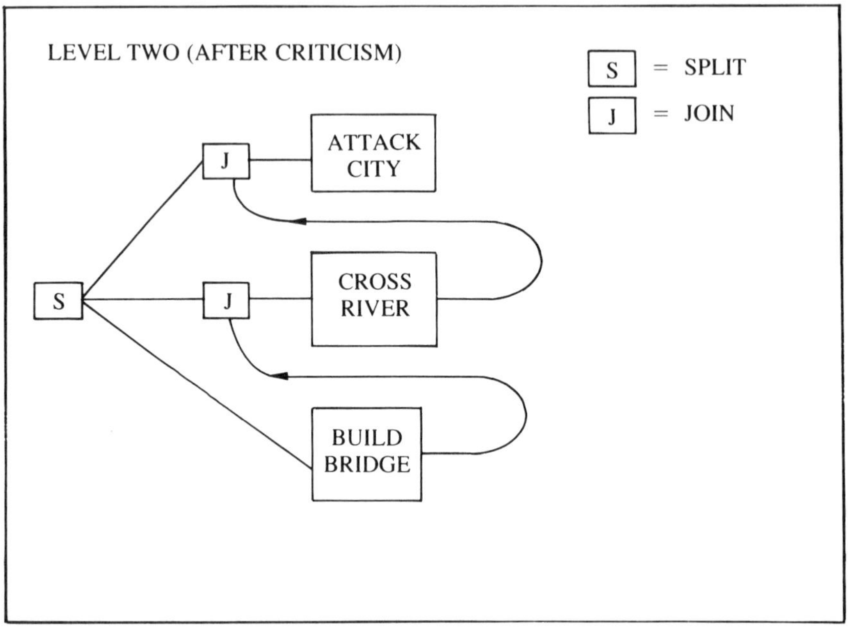

Figure 3-7. Second decomposition level after criticism.

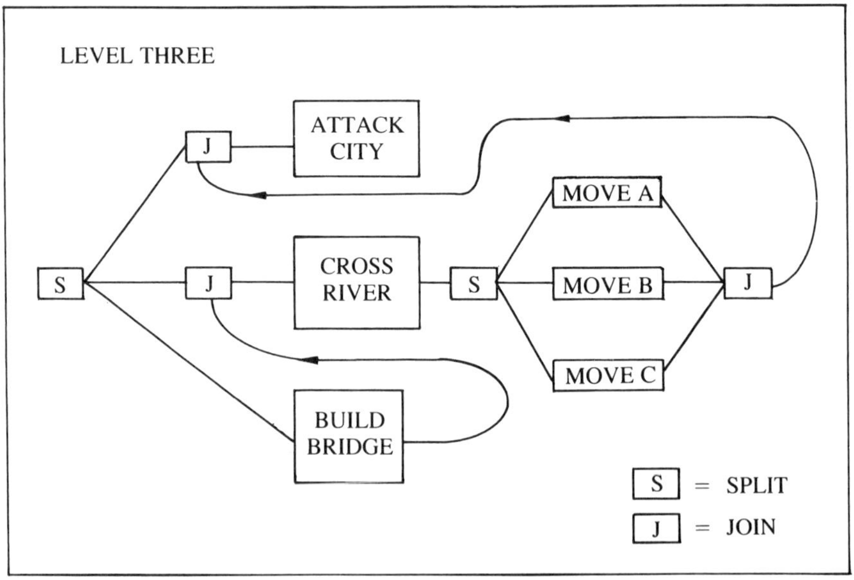

Figure 3-8. Third decomposition level of planning problem.

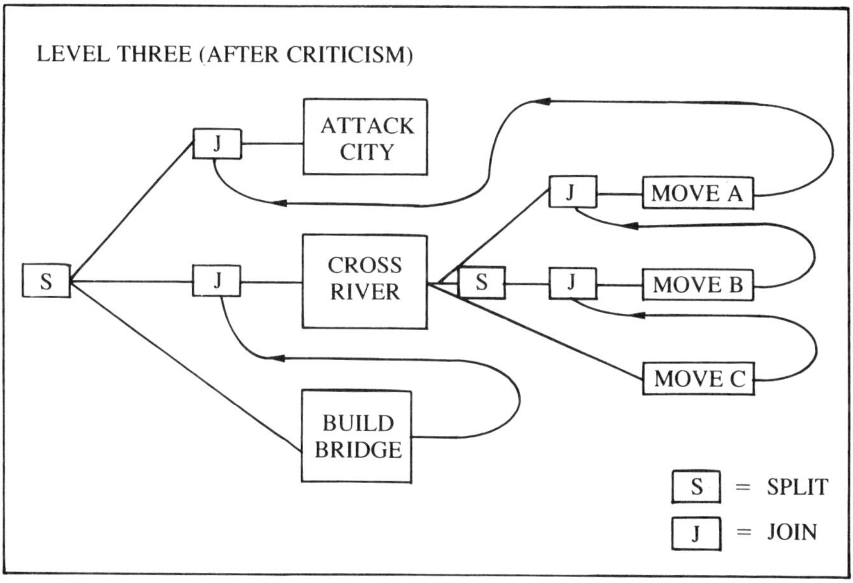

Figure 3-9. Third decomposition level after criticism.

certain range the artillery could be left on the far side of the river to fire on the city from there. A simple resolution to the conflict is shown by the new plan in Figure 3-9.

3.1.4 Skeletal Plans

One interesting fact about human problem-solving procedures which has implications for the development of automatic planning systems is that human problem-solvers typically employ a library of general solution strategies for various kinds of problems. This means that an experienced problem-solver almost never starts completely from scratch in attempting to develop a plan for solving any given problem, but usually has some idea of a general procedure or approach that has worked for a similar problem in the past. This idea has served as a basis for a number of efforts in the development of AI planners, such as MOLGEN, a program which develops plans for scientific experiments based on general experimental design strategies (Friedland, 1979), and PARADISE, a chess-playing program that uses high-level concepts and perceptual patterns to build strategies (Wilkins, 1979). Both systems share an ability to take advantage of what are called *skeletal plans*. A skeletal plan represents a general approach or sequence of steps for solving a set of similar problems. In the CAPTURE_CITY sample plan already discussed, a skeletal plan might consist of the ordered steps of SURROUND_CITY and ASSAULT_CITY. As such, this skeletal plan is applicable to any CAPTURE_CITY problem. Once a skeletal plan has been identified, it is necessary to refine the plan. The refinement process involves fitting the generalized outline of the skeletal plan to the special conditions involved in the actual problem. For example, in the CAPTURE_CITY problem, the BUILD_BRIDGE and CROSS_RIVER subproblems would fall out from the refinement of the SURROUND_CITY skeletal node.

3.1.5 Opportunistic Planning

One of the most appealing approaches to automated plan generation is known as *opportunistic planning* (Hayes-Roth and Hayes-Roth, 1979). In the planning approaches already discussed, the process of plan generation was assumed to be a neat, organized, unidirectional, top-down refinement process. Opportunistic planning makes no such commitment to either an orderly generated network or to a top-down resolution. In fact, the assumptions behind the notion of opportunistic planning are quite contrary to these limitations.

First, the opportunistic paradigm assumes that plan generation can be a multidirectional activity based on the uncovering of opportunities after a particular planning decision has been made. This means, for example, that in the refinement of particular low-level aspects of a plan which are dependent on prior, more abstract parts of the plan, information may be discovered which will suggest a new even more abstract plan that could be used as an alternative to the original abstract plan. To take advantage of such valuable data, a planner cannot work within the framework of a rigid, top-down process; plan generation must in this example be able to proceed bottom-up as well.

Second, earlier planning strategies have been locked into the idea of a hierarchy for purposes of both plan representation and plan generation. Human planners, however, in the process of generating their plans seem to work across hierarchical boundaries. That is, they may make executive-level decisions in conjunction with the refinement of some low-level subproblem, and this may in turn generate an adjustment to some higher-level abstract component of the plan. The planning process may then be thought of as "heterarchically" rather than as hierarchically ordered.

3.2 AI Game Playing

One of the oldest and most interesting traditions within the field of AI involves the development of intelligent computer systems for playing games such as chess, checkers, Go, and many others. Very early in the history of AI, it was hoped that computers would soon be able to play at a level of skill comparable to masters of the games and even beyond. While some significant successes have been achieved in this regard, the reality of computer game playing has proven to be quite different from these initial appraisals. To a large extent, success has been achieved in games such as chess through the development of fast search procedures that can literally examine millions of possible move sequences in a few seconds. In games such as chess or checkers, it is feasible for fast computers to examine all possible move sequences for 8 or more move levels ahead. By comparison, the knowledge-based approach to game playing has been the most successful in games like Go where an exhaustive search to a reasonable level of depth is far beyond today's computers. The knowledge-based approach depends on large amounts of stored patterns, rules, skeletal plans, etc. for evaluating positions and examining a small number of relevant move sequences.

In this section, some techniques in AI game playing are reviewed. This is a selective review that focuses only on techniques relevant to military planning problems. In effect, we are only looking at techniques that might be suitable for playing a war game.

3.2.1 Move Trees

Computer game playing programs are based upon a representation structure known as a *move tree* wherein a particular node represents some arrangement of playing pieces on a board. The "children" of that

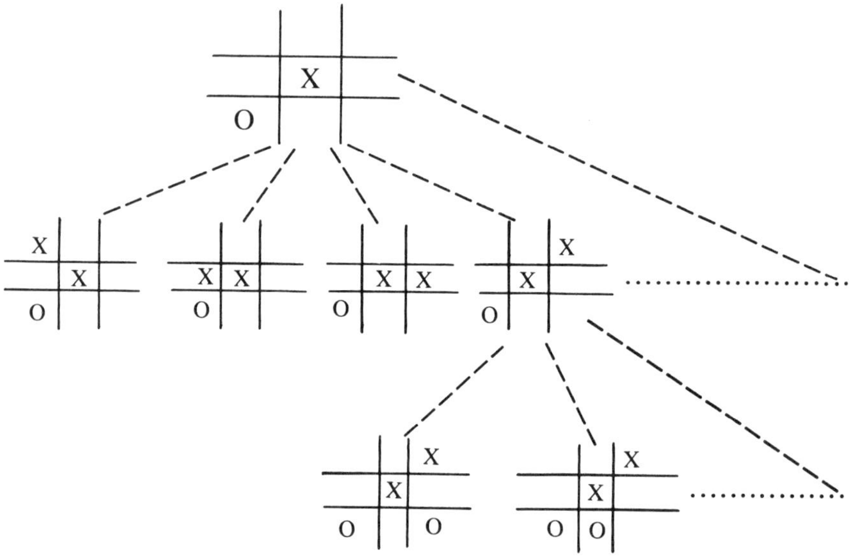

Figure 3-10. Part of a sample move tree.

given node will be the complete set of possible legal moves for the side on the move. Each of these nodes become in turn a starting point for proposing a set of subsequent moves for the opposite side.

Figure 3-10 shows a sample move tree for the game of tic-tac-toe. Reading this tree from the top down we first find all the possible moves for X. At the next level, all possible O responses to X moves are posited. In games such as chess, the classic approach to move tree generation, this expansion can be performed until either the total space of possible moves, countermoves, counter-countermoves, etc. has been generated to some predetermined depth. If an exhaustive search of the possibility space is performed, it is possible to isolate a successful move for Black by identifying a move that will eventually lead to a winning terminal position for Black no matter what response White makes. If the tree generation is stopped after a certain number of levels, then an evaluation procedure must be applied to determine which of the resulting game states represents the "best" situation for Black. An exhaustive search to termination of all possible moves and countermoves is a prohibitively expensive operation in many games. Thus, if a simple tree expansion procedure of examining every possibility is to be used, then the expansion will only be feasible to a certain limited depth of moves.

3.2.2 Tree Pruning

One way to get around the limitations of exhaustive searching is to use *tree pruning*. This means that instead of investigating every possible branch of a move tree, only certain branches are selected for expansion based on the application of a decision procedure at given nodes in the tree. Tree pruning can be accomplished in one of two general ways: algorithmically or heuristically. Algorithmic procedures specify a general mathematical expression that is used to do a numeric evaluation of each position. Positions with the low evaluation are dropped from the search. Heuristic procedures use various types of knowledge elements (patterns, rules, skeletal plans) to identify moves that can potentially lead to obtaining an important goal (e.g., queen capture).

3.2.3 Goal-Driven Planning in Game Playing

The use of long-range planning and goal-based knowledge has been employed in several computer game playing programs with varying degrees of success. These systems include Pitrat's chess program (Pitrat, 1977), Wilkin's PARADISE system for solving tactical problems in chess (Wilkins, 1979), and the University of Michigan Go program (Reitman and Wilcox, 1979; Lehner, 1981).

To illustrate this goal-oriented approach, assume that we have a chess position where one player, White, believes it feasible to find a mating combination. Consequently, White starts with the goal of CAPTURE_KING. To do this, White isolates a square he needs to occupy with his queen to achieve checkmate. However, this square is protected by a Black knight. Consequently, REMOVE_KNIGHT from this square becomes a subgoal. This type of subgoal can be achieved by actually capturing the piece, so CAPTURE_KNIGHT becomes a subgoal of REMOVE_KNIGHT. If the knight can be directly captured, then that capture move becomes the first move. If not, a new subgoal THREATEN_KNIGHT is activated that results in a move that threatens to capture the knight on the next move. In this example, therefore, the first move of a goal-driven look-ahead is one that threatens the knight, so that it is no longer defending the Black king, and so checkmate can be achieved. By carefully considering objectives and goals at each step in the look-ahead in this way, a program can significantly constrain the number of moves/branches to be examined, with the added advantage that these branches will not be subject to the horizon effect.

3.2.4 Goal Pairing

The space of available moves options can be even further constrained if attempts are made to posit goals for the opponent. Surprisingly, this is often not nearly as difficult as might be anticipated because of a general technique (often only implicitly used by developers of knowledge-based game playing programs) referred to as *goal-pairing*. Goal-pairing (Lehner, 1981) is simply the process of identifying an adversary's countergoal to each agent goal or subgoal that has been generated. In our chess example, for instance, it is reasonable to assume that White's goal of CAPTURE_KING corresponds to a black goal of SAVE_KING and, furthermore, that Black would like to prevent the removal of his knight by preventing its capture. Therefore, after a White move that threatens Black's knight, Black's hypothetical response will be a move that removes that threat. Figure 3-11 lists the White/Black goal-pairs that would result from this example.

3.2.5 Representative Search

In many adversarial planning situations, in both game playing and real-life conflicts, human planners consider high-level strategic plans that cannot be analyzed by the same type of precise action-by-action

WHITE GOALS		BLACK COUNTERGOALS
CAPTURE_KING	→	SAVE_KING
↓		↓
REMOVE_KNIGHT	→	PREVENT_REMOVE_KNIGHT
↓		↓
CAPTURE_KNIGHT	→	SAVE_KNIGHT
↓		↓
THREATEN_KNIGHT	→	STOP_THREATEN_KNIGHT
↓		

Figure 3-11. An example of goal pairing.

look-aheads that are represented by a move tree. This is because the strategic plan is at such a high level that there will usually be a large number of reasonable move sequences consistent with high-level strategic goals. Put another way, even after goals and objectives are specified for both sides, there are too many reasonable plans of action consistent with those objectives and goals to be able to consider them all. In military planning, for instance, it is not feasible to translate an Army Corps commander's concept plan into a precise plan of action where all possible friendly and enemy actions are taken into detailed consideration and anticipated. Despite these problems, it is clear that human planners are able to perform some type of strategic planning or look-ahead that anticipate likely results.

What often makes this type of strategic planning possible is that usually all the action sequences that follow the basic pattern in a strategic plan will have a number of common positional side-effects. So, any one sequence of actions that attains a strategic plan is representative, strategically, of all the possible sequences. Consequently, it is possible to determine some of the consequences of a proposed strategic plan by examining just one specific plan of action to implement that plan and to generalize the results to other possible plans of action. In Lehner (1983) this is referred to as *representative search*; in a military context, it is sometimes referred to as "playing out a scenario".

3.3 A Theory of Adversarial Planning

A primary constraint in automated action planning systems is that while general solutions strategies can be applied with admirable success to many problems, these strategies (whether dependent upon metaplans, skeletal plans, critics or other devices) just can't deal with the very dynamic problems that face many military planners. A useful analogy, which epitomizes the problems facing military planners, is the case of most, if not all, present-day robotics. While these machines may be quite adept at piling blocks one on top of another, it is difficult to imagine that they would be successful if impeded by some adversary who kept unpiling and hiding the blocks as fast as our robot got them piled.

Game playing programs seem to suffer from an almost opposite fault. In the absence of long-range plans and goals, these programs tend to "worry" about too much, exploring every possible contingency whether it makes sense or not and thereby wasting time and resources in a pointless search.

Implicit in both domains of automated action planning and computer game playing is the idea that the computer agent is acting in an adversarial world. By "adversarial world" we mean to point out the fact that the difficulties in attaining any goal are the result of attempting to act in a world fraught with various adversities which can include the agent's own inefficiency, other agents (including opponents with their own goals), and hills, rivers and trees that can get in the agent's way.

Recognition that the world is, in a sense, adversarial has two effects. First, based on this observation, we can develop an interpretation that explains the gap between any given world state and some future goal state that an agent would like to achieve. In both action planning and game playing, the basic problem is to close that gap and make the actual world coincide with the goal state the agent has in mind. All planning, whether of moves in a chess game or steps in the attack of a city, is a process of finding the correct steps to force the adversarial world into a desired shape. The second result is that, given this broad and general understanding of adversity, we can proceed to approach automated planning and problem-solving from a fresh viewpoint.

To address the planning problems that involve an intelligent adversary, Young and Lehner (1986) have taken the approach of generalizing the goal tree/procedural network formalisms to represent plans in a manner that permits incorporation of the type of planning done in knowledge-based game playing research. To this basic framework we have added explicit consideration of adversarial countergoals.

The basic premise of this research has been that most of the previous work in AI planning, such as that derived from robot problem-solving, cannot be readily applied to military planning problems. Specifically, these planners lack a satisfactory capability to explicitly incorporate an adversary's goals and actions into the planning process. Consequently, they cannot effectively plan against an adversary that is simultaneously planning against them. As an example, consider the case of planning under conditions of uncertainty. One accepted technique is to include information-seeking goals in the plan whenever information necessary to complete a plan is not initially available to the planner (e.g., FIND_LOCATION (X) if the location of X is not known). When an adversary is present, however, that adversary is likely to have a countergoal of preventing the collection of the required information. Consequently, this adversary will use various tactics, perhaps including deception, to prevent information collection. Unless the adversary's countergoals and actions are explicitly taken into account and planned against, the original information goal is not likely to be achieved.

3.3.1 Contingency Goal Trees

A planner described in Young and Lehner (1986) named ARES (Adversarial REasoning System) represents plans in a structure called a contingency goal tree (CGT). Figure 3-12 is a sample CGT showing a contingency plan for a hypothetical battle situation. At each node in a CGT is a goal pair (GP) which includes a proposed friendly goal and a possible countergoal for the adversary. Subnodes of GPs are subgoal pairs (SGPs). Reading the leftmost branch of our example CGT, we see that a subgoal of CAPTURE(A) is SURROUND(A), while the corresponding subgoal of DEFEND(A) is RETREAT(A). Consequently, the CAPTURE(A):DEFEND(A) has as its first SGP SURROUND(A): RETREAT(A).

Simultaneous multiple tasks are treated as conjunctive (AND) and disjunctive (OR) goals. Thus (AND ENC(P1) ENC(P2)) is a single goal of enclosing both escape path 1 and escape path 2, while (OR ESC(P1) ESC(P2)) is a goal of escaping through either path 1 or path 2. Other "generic" goals such as AND_IN_SEQUENCE are possible.

In many instances, GPs will include a goal or countergoal of NIL. This occurs when there is no projected opposing goal, such as when ARES is exploring an option where one side pursues an independent course of action while ignoring the adversary's goals. It also happens to occur in our example CGT because this example depicts a linear sequence of moves and countermoves, such as would be found in alternating move games like chess.

When a CGT contains a NIL element in every GP, it is isomorphic to a standard goal tree. Also, CGTs can have parallel branches and, if desired, preconditions and post-conditions attached to nodes. Consequently, CGTs represent a straightforward extension of the goal tree/ procedural network representation used in action planning.

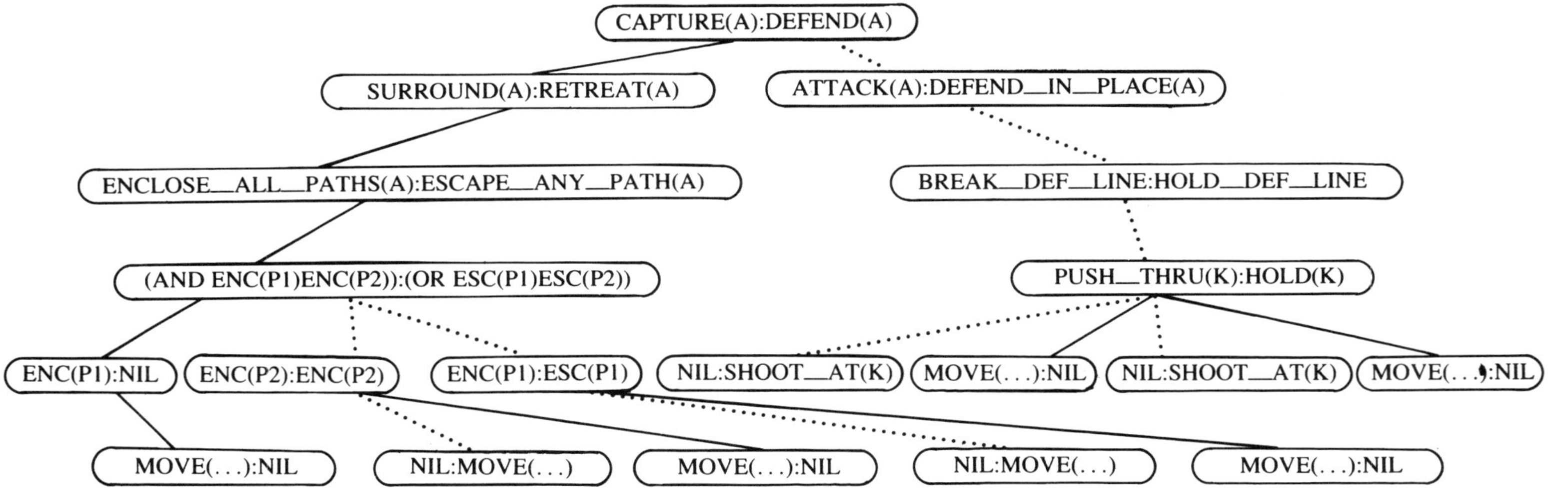

LEGEND			
	GOAL:COUNTERGOAL PAIR	A	TARGET (e.g., an infantry unit)
(AND. . .	CONJUNCTION	P1	PATH
(OR. . .	DISJUNCTION	P2	PATH
ENC	ENCLOSE	K	POINT ON A DEFENSIVE LINE
ESC	ESCAPE		

Figure 3-12. A sample contingency goal tree (CGT).

3.3.2 Plan Generation in Adversarial Environments

By reading the terminal nodes of a CGT from left to right, a specific sequence of actions is defined. Consequently, each CGT specifies a contingency plan of what to do if the adversary pursues the goals indicated in the tree. In an adversarial world, "good planning" means generating contingency plans for each reasonable course of action that either the adversary or agent may pursue. To this end, ARES attempts to generate a set of CGTs that provide contingency plans for the full scope of reasonable options available to the agent and the adversary.

For a given level of abstraction, ARES proceeds through CGT generation/expansion in a depth-first manner. Beginning with the most recent GP added to the CGT, ARES recursively adds SGPs until a terminal GP is reached. A terminal GP is distinguished by the fact that it includes a procedure for making an action in the world, which means that it is the lowest level in the expansion. Processing of a GP involves accessing a knowledge base of goal definitions to determine possible reasonable options to instantiate for either the agent or adversary. In alternating move games, these accesses are from the goal for the side that is currently on the move. In more complicated domains like maneuver planning, where action can occur simultaneously, both parts of the GP are processed at the same time. For an in-depth treatment of this approach in ARES see Lehner (1988).

The general procedure employed by ARES in the generation of plans may be characterized as an iterative process of proposing possible courses of action and, based on the outcome of those actions, proposing better action options. ARES in effect plays out a series of hypothetical games in order to "learn" about the way objects in the domain interact. The knowledge that is gained is further used in the construction of the the network of goals and countergoals.

3.4 Summary

In the last few years, there has been something of a Renaissance of research interest in automated planning. This renewed interest is to a large extent the result of the recognition that planning is a key component of many aspects of military decision-making. While this interest is encouraging, it should be recognized that very few automated planning systems for military decision problems have even made it to the completed prototype stage.

Examples of military planning system prototypes are found in Chapter 8 where some of the difficulties in developing automated planning systems for our national defense are discussed.

4
Basic Concepts in Computer Vision

Computer vision is a complex and diverse research area oriented toward the processing and interpretation of two-dimensional images. For our purposes, an image can be defined as a two-dimensional function $f(x,y)$ that specifies the value of each point (x,y) in an image. Each (x,y) point in the image is called a *pixel*. The range of the $f(x,y)$ function differs with the type of image. In *monochromatic* black-and-white pictures, the function specifies the level of gray intensity. For panchromatic color images, the function describes a vector of color intensities. In some cases, the function can also specify a distance value for each point. The task of a computer vision system is to "understand" the scene depicted in an image; that is, to identify the objects embedded within the image.

Computer vision research is not unique to AI. Several research fields claim computer vision to be within their purview of interest. While the techniques from these research fields overlap, frequently borrowing from each other, each has its own orientation and goals.

One category of vision research techniques (from Cohen and Feigenbaum, 1982) involves signal processing, image classification, and image understanding. These are described by Cohen and Feigenbaum (pp. 127-128) as follows:

- *Signal processing*. Signal processors transform an input image into another image that has desirable properties. For example, the output image may have a better signal-to-noise ratio or may be enhanced by emphasizing the details to facilitate human inspection. The content of the image is often irrelevant. *Image processing* and *picture processing* are the most common terms for this class of processing. As well as digital techniques, optical techniques and electric video-signal techniques can often provide very fast throughput.
- *Classification*. Classification techniques classify images into predetermined categories. Character recognition is a typical example. Often, a predetermined set of future values is extracted from images, and the decision of how closely an image "fits" a class is made on the basis of statistical decision methods applied to the multidimensional feature space. There is a large body of theory for designing optimal decision rules. These methods are usually called *pattern recognition* or *pattern classification*, although the word "recognition" is used only for historical reasons.
- *Understanding*. Given an image, an *image-understanding program* builds a description not only of the image itself but also of the scene it depicts. In AI vision research, the term *scene analysis* was often used to emphasize the distinction between processing two-dimensional images (as in pattern classification) and three-dimensional scenes. Image understanding requires knowledge about the task world, as well as sophisticated image-processing techniques.

AI has focused primarily on the problem of machine understanding of images, and has consequently borrowed from the other two classes of techniques as needed to assist the understanding process. A brief overview of image understanding technology is provided below.

4.1 An Overview of Image Understanding Techniques

Broadly described, image understanding involves two basic steps: *segmentation* and *interpretation*. *Image segmentation* is the process of translating a "raw" image into an intermediate representation that describes the lines and regions within that image. *Image interpretation* is the process of examining the regions in a segmented image to identify the objects in that image.

Image segmentation is the process of extracting lines, edges and regions from an image. Although a large number of very specific techniques have been developed to support this process, all are based on a search for adjacent pixels that have either contrasting or similar values. *Edge detection* generally proceeds as follows. At each pixel location, the local area around that pixel is explored to determine whether it may be an edge. It is classified as a possible edge if there is a sharp difference between the value of the pixel and its neighbors. Assessment of a "sharp difference" can be done by computing the derivative, or by comparing the local pattern of pixel intensity with that of an edge template.

Once edges have been identified, they must be grouped into meaningful lines. *Line detection* is done by tracing the detected edges according to some set of rules. The edges of an image generally have gaps, sudden curves, multiple forks, false edges, etc. As a result, it is often difficult to follow a meaningful line, and the corresponding set of rules for handling these difficulties may be quite complex.

Once lines and edges have been found, the image must then be segmented into *regions*. Regions are areas, surrounded by detected edges, that have pixels of similar values. The objective of the segmentation process is to translate an image into a set of separate regions that correspond to the objects and object-derived features (e.g., shadows) in the scene depicted by the image. Higher-level image understanding and object identification can then proceed. Unfortunately, despite a great deal of research in this area, segmentation procedures will generally only produce a crude approximation to the true regions of an image. For instance, to detect all the edges in a region, a low threshold is used, but this will often result in a number of false alarms of spurious edges also being detected. Line-tracing algorithms will often separate a single line into two separate lines or merge two distinct lines into one. As a result of errors such as these, the regions defined in an image will be either over- or under-divided. The surface of an object in the scene will often be divided into multiple regions in the segmented image, while at the same time, distinct surfaces from several objects may be merged into a single region.

Figure 4-1 shows a typical example of a segmented image. As can be seen, trying to identify the objects based only on the information in Figure 4-1 is far from trivial, yet this is exactly what image understanding algorithms must do. The image in the figure was a picture of an urban area complete with numerous buildings and city streets.

Once a segmented image has been generated, the computer vision system must interpret it. The approach generally taken is to examine the regions in the image to determine whether they match features from one of a predetermined set of possible objects. In effect, to limit the search space of possible interpretations, AI vision researchers have relied heavily on the development of models that guide what the image interpretation mechanism expects to see.

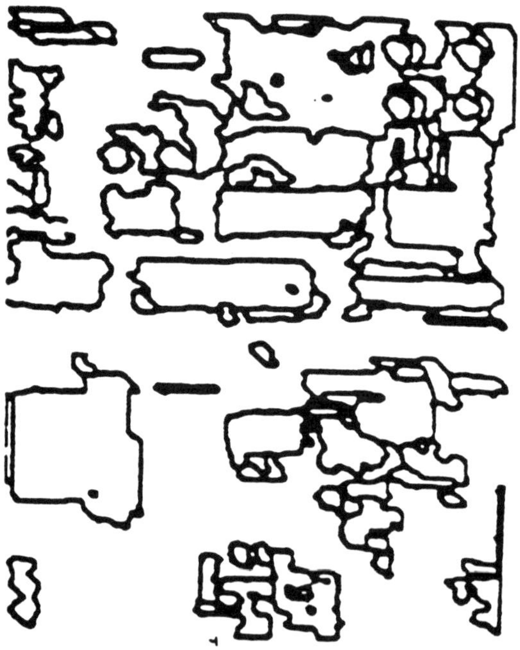

Figure 4-1. A computer-segmented image.

Developing models to guide expectancies has an added benefit. Namely, if the vision system has some idea as to who it expects to see, then the parameters in the segmentation process can be tuned to maximize the probability of detecting the edges, lines and regions of the expected object. As a result, not only can the interpretation process be controlled by a top-down model, but the low-level image processing techniques can be controlled as well.

4.2 Two Examples of Vision Systems

Provided below are two examples of vision systems. The first provides an instance of how a vision system can exploit world knowledge to focus visual information processing. The second is an example of a general system that can be adapted to a variety of image understanding problems.

4.2.1 Using World Knowledge to Constrain Search

As noted above, if a vision system contains a model of the world it is trying to perceive, then it is possible to use that knowledge to control the image interpretation process by looking for expected features. A classic example of this approach is found in Waltz's algorithm (Waltz, 1975) for interpreting simple blocks world images. Consider the line drawing in Figure 4-2. Drawings such as this reflect lines that may be obtained from the segmentation of an image of several blocks. The goal of Waltz's algorithm is to identify the objects in that image, viz., the blocks.

To interpret images, such as in Figure 4-2, Waltz performed a detailed investigation of the possible line types that might bound an object in a line drawing. If one ignores shadows and object imperfections, there are only three types of lines possible. These include:

- Obscuring edges that reflect a boundary between an object and another object or background;
- Concave edges that form an acute angle between two faces; and
- Convex edges that form an obtuse angle between two faces.

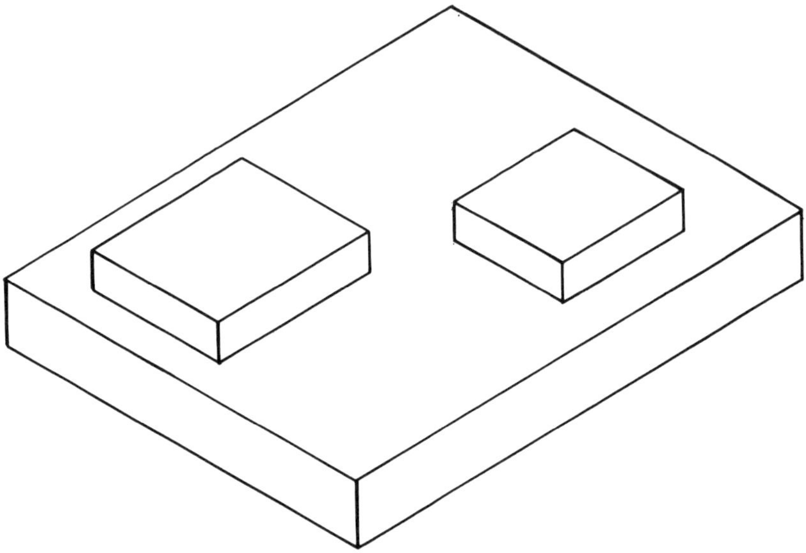

Figure 4-2. A simple segmented blocks world image.

The first step in interpreting a line drawing is to determine for each line in the drawing exactly which type of edge it is. Applying the labeling convention shown in Figure 4-3 to our sample drawing, the labeling in that figure results. Once this labeling is performed, it is a straightforward task to discriminate the block in an image. The goal of Waltz's program is to perform this labeling.

The key to Walz's approach is the recognition of the fact that in the blocks world only a small number of possible vertice types are possible. If we examine domains where there are only trihedral vertices (vertices that join exactly three planes), then only four types of vertices are possible. These are shown in Figure 4-4. In principle, there are 208 possible patterns of line labelings that can result from these vertices, of which only 18 are physical possibilities. These 18 are shown in Figure 4-5.

Using the knowledge that only 18 distinct vertex labelings are possible, and that each line must be uniquely labeled, the problem of image interpretation and object extraction becomes a straightforward search process. Specifically, the goal of the computer vision system is to find a labeling of the lines in the drawing such that each vertex has a pattern of labels that maps into one of the 18 patterns. By looking sequentially at

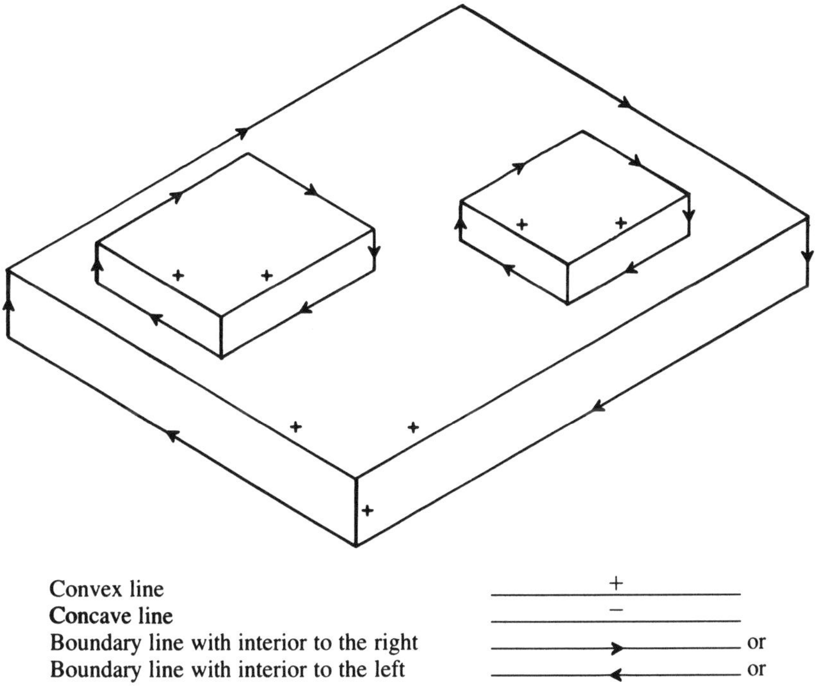

Figure 4-3. An example of line labeling.

connected vertices (like connecting the dots), it is possible to quickly narrow down the possible interpretation of each new vertex examined.

Overall, Waltz's approach is a good example of how world knowledge can be used to limit the number of features a vision system needs to search, and as a result to quickly constrain the process of searching alternative possible interpretations of an image.

4.2.2 A Domain-Independent Vision System

The program described above is highly domain dependent. That is, the program is only designed to solve problems within the blocks world domain. It exploits properties of that domain that may not be invariant for problems outside that domain.

A more domain-independent system is one that possesses a variety of procedures for exploiting perceptual capabilities, but allows the user to

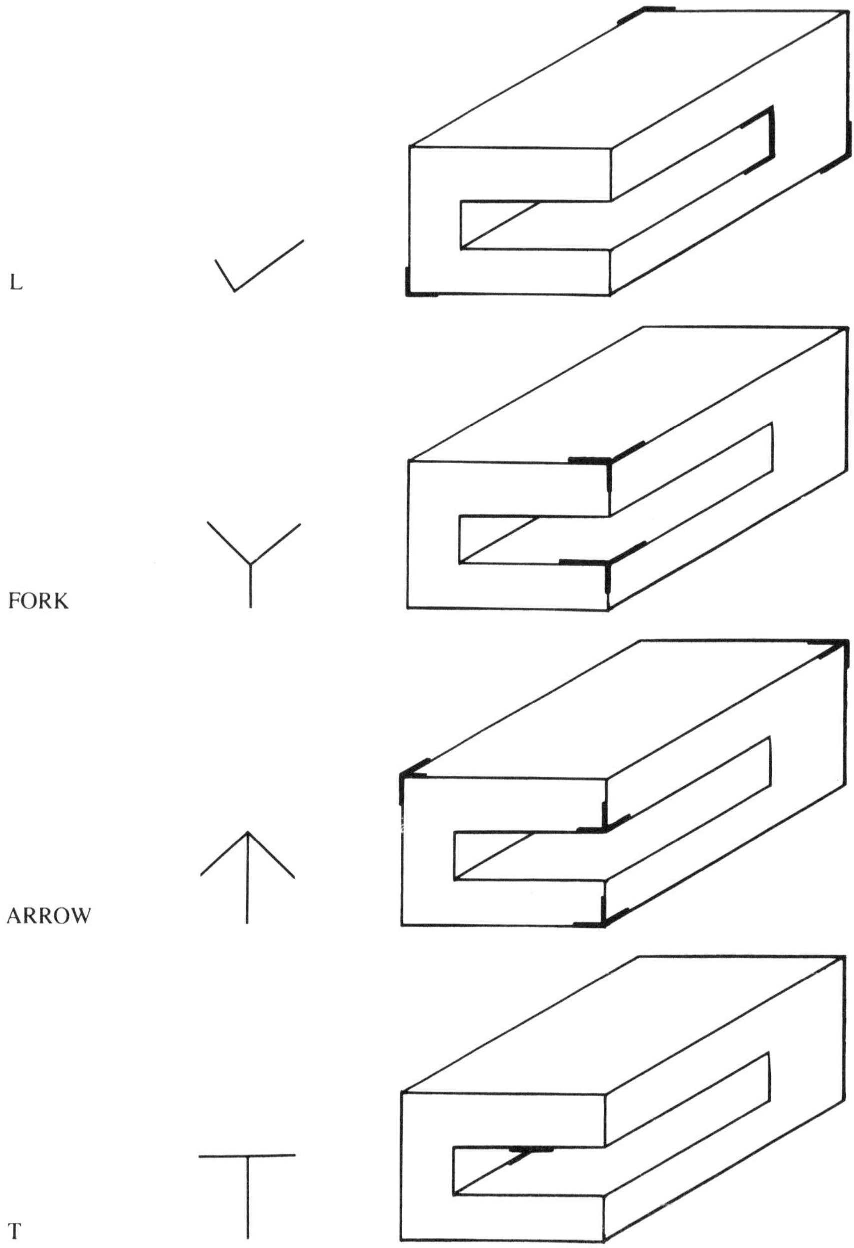

Figure 4-4. The four types of trihedral vertices.

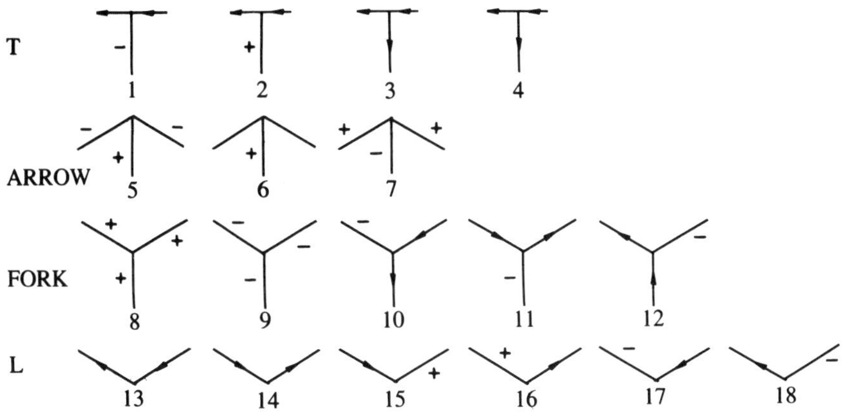

Figure 4-5. The 18 physically possible trihedral vertices.

enter a world model; this world model represents a knowledge base of the objects found in a problem domain and the characteristics of the objects that may be found in a given domain. In this type of system, the world model is used to make predictions about what is likely to be observed, and then the perceptual capabilities focus on finding these anticipated observables. This process is iterative in that, as the perceptual system reports observations, these observations are matched against the present predictions and world model, and new predictions are made. This general process is shown in Figure 4-6.

Perhaps the best-known example of a system that allows the user to specify a world model is ACRONYM. ACRONYM is considered to be

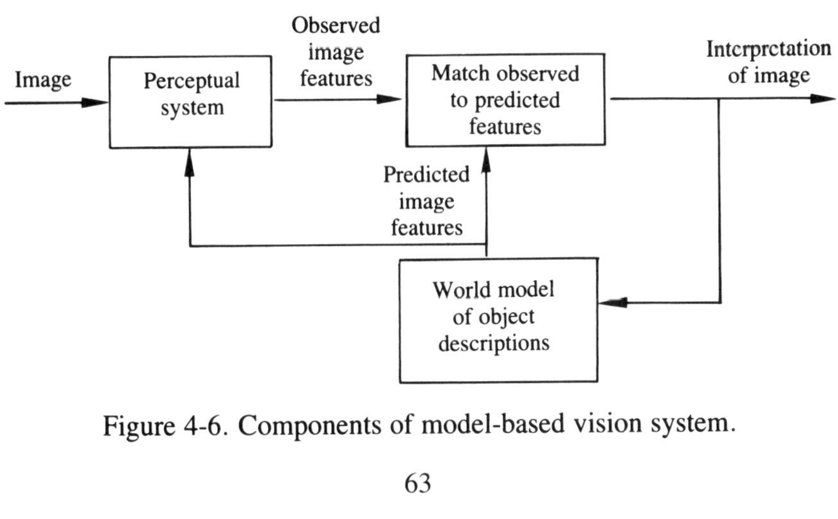

Figure 4-6. Components of model-based vision system.

a model-based vision system that decomposes vision into four parts: modeling, prediction, description, and interpretation (Brooks, 1981). Each is briefly described below.

Modeling

ACRONYM provides users with an object description language that allows them to characterize three-dimensional objects in a domain of terms of (1) geometric properties such as shape and volume, (2) component subobjects within an object, (3) spatial relationships among these components, and (4) a series of algebraic expressions that allow a user to parameterize many of those properties and relationships. The object descriptions are represented in ACRONYM by an *object graph*, while algebraic relationships among components (e.g., relative size) are represented in an associated structure called a *restriction graph*.

Prediction

Given a set of objects models and their interrelationships, ACRONYM applies some internal geometric analysis procedures to identify features of objects that will be invariantly observable if the object is present in an image. For the class of objects defined in an object model, ACRONYM attempts to extract a sufficient number of invariant features to uniquely discriminate objects in one class from other possible objects in a domain and to determine the position and orientation of an identified object. ACRONYM does not predict all features that may be observed as part of an object.

The product of this prediction process is a *prediction graph*. The nodes in a prediction graph are the object features that should be observable, while the arcs characterize the relationships among these features (distance, relative size, connectivity, etc.).

Description

For a given image, ACRONYM applies some line-finding and segmentation procedures to identify a set of regions in a two-dimensional image. These regions are characterized in terms of their shape and relative location and orientation in the image. Region descriptions are maintained in a structure referred to as a *picture graph*. The picture graph characterizes the elements of an image in terms of the same features and variables as found in a prediction graph.

Interpretation

In ACRONYM, interpretation is accomplished by matching the picture graph description to a prediction graph. If there is a reasonable match between the prediction graph and the picture graph, then the object is identified.

4.3 Summary

A considerable amount of military intelligence is based on photographic and other types of images. The range and variety of image collection capabilities that exist today can result in a volume of diversity of image interpretation problems that far exceed the capacity of existing systems, manual or otherwise. As a result, there is a natural interest in the development of automated image interpretation capabilities.

With the advent of relatively domain-independent systems such as ACRONYM, it is becoming feasible to develop generic image understanding systems that can be applied to a variety of problem domains. ACRONYM, however, was an experimental system. Work in this area has continued and has moved into the development of prototypes of operational systems. A discussion of such a system is provided in Chapter 8.

5
Other Types of AI Systems

Most applications of AI technology to problems in national defense involve the development of an expert system, natural language processing system, automated planning system and/or an automated image understanding system. Chapters 1 through 4 provide a general introduction to the basic types of AI systems that are of most interest to the national defense community and from the perspective of this community show the most potential for practical applications in the near term.

There exist, however, many other classes of AI systems and techniques that provide significant potential opportunities for national defense applications, but have so far received comparatively little attention. Some of these systems are briefly overviewed in this chapter.

5.1 Intelligent Training Systems

A general application area of AI techniques is in the development of intelligent systems for training and education. Training is a major expense in the military. As noted in Fletcher (1985), on an average day in 1983 about 214,000 active duty personnel, 40,000 National Guardsmen and Reservists, and 12,000 foreign military students and civilians were undergoing some type of formal training. The development of any techniques that could streamline or reduce the expense and time required for training would be of significant value. The applicability of AI to education and training problems are discussed below.

5.1.1 Intelligent Computer-Aided Instruction

Computer-aided instruction (CAI) is an educational technology oriented toward developing instructional programs that provide for the individualized tutoring of well-prepared course materials. While many CAI programs have been developed, most are limited to the presentation of standard instructional materials, or, if students consistently repeat certain errors, the presentation of standard remedial materials. As a result, the only aspect of instruction that is tailored to individual students is the order of presentation of the instructional materials and not the instruction itself. This limitation is an inherent result of the fact that CAI programs do not themselves have any knowledge about the domain in which they are providing instruction. This is unlike a human teacher who will understand (hopefully) the problem domain he or she is teaching. A good teacher, upon observing that a student is making errors, will engage in several cognitive processes that will result in providing the student with proper instruction. These processes include the following:

1. Determine the correct answer.
2. Think about the process required to generate the correct answer.
3. Identify the knowledge required to execute this process.
4. Hypothesize about what aspects of this knowledge the student

is lacking. (This may involve trying to vicariously simulate the student's thinking.)

5. Select a strategy for instructing the student as to the required knowledge.

6. Instantiate the selected strategy with specific instructional materials that the teacher may generate on the spot.

Obviously, a teacher's cognitive processes for tailoring instructional materials to the student are considerably more sophisticated than the simple order and branching techniques found in CAI devices.

Intelligent computer-aided instruction (ICAI) is a research area devoted to the development of instructional systems that have a true knowledge base of their teaching domain. That is, they can, in principle, intelligently solve the same problems that they are trying to teach students to solve.

At first, it might appear that building an intelligent instructional system for a domain is a straightforward extension of the problem of building an intelligent system for solving problems in that domain. After all, if a medical diagnosis expert system can be built that emulates the reasoning of human doctors, it should be relatively easy to extend that system to teach medical diagnosis techniques. This turns out to be quite false. MYCIN, for instance, is an excellent program for diagnosing bacterial infections, but is not very useless as an instructional system even when adapted for this purpose (Clancey, 1984). The reason has to do with the types of knowledge that are required for proper instruction. These are discussed below.

Components of an ICAI System

To emulate some of the techniques used by a good instructor, ICAI systems must have at least three separate "intelligent" components: a domain knowledge base, an approach to student modeling, and a teaching strategy. These are discussed below.

Domain knowledge base

This component contains all of the knowledge in the ICAI system relevant to problem-solving in the instructional domain. This component is responsible for generating test problems and evaluating the correctness of the student's solutions to the problems. This component should be sufficiently expert in its own right to be able to solve the test problems.

To obtain the required problem-solving expertise, the knowledge base component of an ICAI system will generally reflect the same knowledge representation schemes found in other AI-based problem-solving systems. Semantic networks are sometimes used to represent facts, causation and relationships, scripts may be used for representing causality and temporal relationships, production rules are used to emulate problem-solving heuristics used by human domain experts, etc.

The knowledge-based expert component of an ICAI system is sometimes referred to as an *articulate expert* if it can explain the decision processes necessary to solve a problem in a way that is consistent with human problem-solving processes. Such articulation can be very difficult to program even in systems that emulate human problem-solving processes. The MYCIN system, for example, diagnoses and recommends treatment for bacterial infections using production rules that correspond to the same heuristics used by medical doctors. In research performed by Clancey (1984), it was discovered that the standard rule trace used by MYCIN to explain its diagnostic reasoning to doctors was not very useful for instructing diagnostic reasoning to students. Students indicated that the MYCIN rules were difficult to understand and remember and did not incorporate easily into their problem-solving processes.

In order to convert the MYCIN knowledge base into an articulate expert knowledge base, several enhancements and additions to the knowledge base were required. First, the control strategy for choosing and ordering which diagnostic rules to test at any given time had to be explained to students. (That is, MYCIN has embedded within its control procedures a general approach to diagnostic problem-solving. Since this problem-solving strategy was encoded into the MYCIN program, it was never declared explicitly in the rule base. Therefore, MYCIN was not able to articulate it to the students.)

Second, the MYCIN rules themselves were not self-explaining. These rules, in fact, were essentially summaries of a considerable amount of background knowledge and experience. In order to justify or explain MYCIN rules to students, elaborate justifications of the rules were required. Such justifications were needed not only to provide students with an understanding of why each rule exists, but also to provide necessary mnemonics for remembering the rule.

Third, MYCIN itself has no knowledge of the boundaries of its own knowledge base. That is, it contains no rules for differentiating conditions under which its diagnostic rules are appropriate from conditions under which they are irrelevant (e.g., it will never recommend that something other than infectious disease may be the cause of the problem). Consequently, knowledge needs to be added so as to instruct students as to the conditions under which what they are learning should or should not be applied.

Student models

The objective of an ICAI system is to adapt to the instructional requirements of individual students. To achieve this objective, the system must have some knowledge of the status of the student's knowledge base. The purpose of the student modeling component of an ICAI device is to generate a representation of the student's understanding of the material being taught. The student model component is usually oriented toward generating and testing hypotheses that account for misconceptions and noncriterion performance in students.

The student modeling component can examine several sources of data to generate hypotheses about student misconceptions, including the student's (1) present problem-solving behavior, (2) historical problem-solving behavior, (3) responses to direct queries to the student about what he or she understands, and (4) a characterization of the learning material itself, such as its overall difficulty. Against this evidence, standard pattern recognition techniques can be employed to determine if the student's pattern of errors is consistent with a specific hypothesis regarding his lack of understanding.

The student model itself is a representation of the student's knowledge base. One approach to student modeling is the use of *overlay models* (Carr and Goldstein, 1977). In this approach, perceived student knowledge is matched against the domain knowledge base, and areas of student understanding are flagged. For instance, if the ICAI device has a knowledge base in the form of a semantic network, then individual arcs in the network are flagged as the student exhibits an understanding of the relationship involved. On the other hand, if a production system is used, then individual production can be flagged whenever the student exhibits problem-solving behavior that employs them.

An inherent characteristic of the overlay model is the implicit assumption that the student's knowledge base is a direct subset of the machine's knowledge base. That is, student errors are interpreted only as a *lack* of knowledge; the possibility that the student may have incorrect knowledge is not considered. As a result, an ICAI system using an overlay model will only provide instructional materials that attempt to fill in knowledge gaps. Correcting a student's incorrect ideas can only be an unplanned-for by-product.

Another approach to representing a student's knowledge base is in terms of a set of *bugs*, where a bug is defined as a deviation from his/her knowledge base. The bug model, while theoretically more complete, is also generally more difficult to implement than the overlay model. This is because the system must represent and detect knowledge structures other than the expert knowledge embedded in the ICAI knowledge base.

Tutoring approach

Once an ICAI system has diagnosed a student's errors in terms of incorrect or missing knowledge, it must still provide instruction to correct his knowledge base. There are a variety of approaches/strategies to teaching which significantly differ in the type of instructional materials that they present and the procedures used to select what next to teach.

One approach is called the *Socratic method*. This method involves querying the student about his answers in a way that will encourage him to reason further about his own knowledge, thereby clearing up misconseptions. Shown in Figure 5-1 is a sample interaction with the program WHY which teaches students about meteorology (Stevens, et al., 1978).

Another approach is that of *coaching*. In this approach, the student is involved in some problem-solving or skill-acquisition activity, such as in computer games. The system observes the student's behavior and at selected points interrupts his/her activity to provide new information or suggest new strategies. A coaching program must therefore identify and interpret student errors, select advantageous moments to interrupt the student's stream of activity, and provide advice that results in improved student problem-solving performance.

Summary of ICAI Techniques

ICAI systems provide the potential for intelligent systems that provide adaptive and effective instruction to students. As such, they represent a discipline of research that in the long run may greatly benefit instructional elements of a national defense system through both cost reduction

Do you think the Amazon jungle has heavy rainfall or light rainfall?

** Heavy rainfall

Why does the Amazon jungle have heavy rainfall?

** Because it's near the Atlantic

Yes, the Amazon jungle is near the equatorial Atlantic.
How does that affect rainfall in the Amazon jungle?

** The water in the current is warm

.
.
.

What happens after the mositure-filled air reaches the land?

** I don't know

Do you think the Andes mountain range has any effect on the rainfall in the Amazon jungle?

** Yes

How?

** The prevailing wind hits the Andes and the air cools

Do you think that any place with mountains has heavy rainfall?

** Yes

Southern California has mountains.
Why doesn't Southern California have heavy rainfall?

Figure 5-1. Example of Socratic method from program WHY. Student responses are preceded by two asterisks.

and better instruction. However, this area remains largely a research topic.

Table 5-1, from Fletcher (1985), lists some recent ICAI systems. As can be seen in this figure, the instructional domains of these systems are very specialized and are of little relevance to national defense. The best potential in the near future for the development of practical AI-based instructional systems appears to be in the area of maintenance diagnosis and repair (Fletcher and Psotka, 1986).

System	Subject Matter	Knowledge Base	Student Model	Tutor Model
SCHOLAR	South American geography	Semantic network	Overlay with importance weights	Socratic dialogue management
WHY	Causes of rainfall	Scripts	Misconception identifier	Socratic dialogue management
INTEGRATE	Symbolic integration	Self improving rule-based representation	Overlay	Reactive environment with advisor
SOPHIE	Electronic trouble-shooting	Semantic network with circuit simulator	Overlay	Reactive environment with guided interactions
WEST	Arithmetic expressions	Rule-based representation	Overlay	Reactive environment with coaching
BUGGY	Subtraction	Procedural network	Misconception identifier	Reactive environment with advisor
WUSOR	Logical relations	Genetic graph (network)	Overlay with familiarity weights	Reactive environment with coaching
EXCHECK	Logic and set theory	Rule-based representation with logic interpreter	Overlay	Reactive environment with advisor
BIP	Programming in BASIC	Rule-based representation	Overlay	Reactive environment with curriculum net and advisor
SPADE	Programming in LOGO	Rule-based representation	Overlay	Reactive environment with coaching

Table 5-1. Some ICAI programs.

System	Subject Matter	Knowledge Base	Student Model	Tutor Model
ALBEGRA	Applied algebra	Rule-based representation	Overlay	Reactive environment with coaching
LMS	Algebraic procedures	Rule and rule-based representation	Rule-based diagnostic representation	Reactive environment
QUADRATIC	Quadratic equations	Self-improving rule-based representation	Overlay	Reactive environment with advisor
GUIDON	Infectious diseases	Rule-based representation and meta-knowledge	Overlay with application probability	Reactive environment with structural interactions
MENO	Programming in PASCAL	Semantic network	Misconception identification	Reactive environment with advisor
STEAMER	Steamship propulsion	Device model	Overlay onto device model	Overlay onto environment with advisor

Table 5-1. Some ICAI programs (continued).

5.1.2 Other Types of Training Systems

In addition to ICAI applications, AI techniques can benefit training in other ways. Two that are frequently discussed is the use of AI for embedded training and the development of intelligent simulations. *Embedded training* refers to the development of training modules that could be incorporated into operations systems (e.g., tank artillery, communication systems, etc.). These modules would provide instructions to users of these systems on how to use the system, as well as on the problem domain for which the systems were developed. Embedded training systems would incorporate many of the ICAI techniques discussed above. *Intelligent simulation systems* would provide for simulation capabilities that would be dynamically adapted to a simulation as it unfolds. For instance, in a battle simulation, an automated adversary could "play red" and dynamically respond to the actions of the simulation user.

5.2 Machine Learning Systems

The concepts of intelligence and learning are highly intertwined. Indeed, in dictionary definitions, "intelligence" is often associated with the ability to understand (learn) things quickly or to perform other forms of *adaptive* behavior. Unfortunately, while the connection between intelligence and learning is quite reasonable, the ability to learn is not characteristic of most AI systems today. The knowledge base of an expert system, for instance, is usually a static structure that is only changed by a programmer or knowledge engineer. As a result, knowledge base development is a very time-consuming and labor-intensive task.

Machine learning remains very much a research field in AI, with relatively little to show in the way of applications. However, as an area of research it is very active (see Michalski, et al., 1983, 1986). A number of scientists are exploring automated and semiautomated techniques for developing a knowledge base that could eventually lead to the rapid development of expert systems and other types of AI systems.

5.3 Automatic Programming Systems

Automatic programming is the automated generation of software code. The goal of automated programming in AI is to allow nonprogrammers to generate code by having the machine translate user-specified functions into proper software that can execute those functions. It is not presumed a priori that user-functions are precise or well-defined. Rather, user specifications will be in a natural language or other imprecise form. Consequently, AI-based automatic programming systems generally work interactively with users to iterate through a series of software programs that represent successively better approximations of what the user wants coded.

This approach to automatic programming is somewhat different than that suggested by Martin (1982) and others where precise, graphic-oriented techniques are used to specify functions and to then automatically compile these well-defined requirements into code.

For the most part, AI-based automatic programming techniques are not widely used. One exception is in the development of natural language interfaces (NLI) where the goal of the NLI system is to provide an interface to one or more data base systems. Generally, a NLI will generate a data base retrieval program as the product of its natural language processing of user inputs (see examples in Chapters 6 and 7). Data base retrieval is relatively simple, and is oriented toward one-time applications. That is, once the data retrieval is successful, the automatically generated program can be discarded.

Other areas of automatic programming involve the automatic development of software that will be used repeatedly, and consequently may need to meet stringent performance specifications (e.g., reliability, validity, processing speed, etc.). Research applications involve the development of repeated-use programs in such areas as robot planning (Sussman, 1975) and geological analysis (Barstow, 1984).

While automatic programming is not a major aspect of ongoing AI applications in national defense, it has been discussed as a possible key element of the Strategic Defense Initiative. (See Chapter 10 for a discussion.)

5.4 Nonmonotonic Reasoning Systems

AI systems for making deductions (e.g., expert systems) operate by accepting inputs on the status of primitives, and then performing some reasoning process based on those primitives to generate conclusions. Since absolute certainty about the status of primitives is rarely the case, estimates, guesses and assumptions must often be made about them. Consequently, a deduction may be based on any number of assumptions that may later be shown to be false. When this occurs, most AI systems can do little but restart the reasoning processes from scratch to determine if the previously generated conclusions are still valid.

Nonmonotonic reasoning refers to a logic for dealing with reasoning under assumptions (Doyle, 1979). Nonmonotonic reasoning systems provide a deduction mechanism that keeps track of assumptions and any intermediate or final conclusions based on those assumptions. Whenever an assumption proves false (or is simply no longer believed), the system will identify all consequences of that change. These consequences may come in a variety of forms: previous conclusions may now prove false, other conclusions that were initially based on the no-longer-accepted assumption may still prove true because of other logic chains, still other assumptions may be rejected by consequence (e.g., if A and B imply C, and C is known to be false and B true, then A must be false), etc. Nonmonotonic reasoning systems can operate either as independent systems or as adjuncts to other reasoning systems (e.g., an expert system using some other logic) to keep track of multiple paths to a conclusion.

While nonmonotonic reasoning will eventually be an important aspect of many AI systems, it is not now a major mechanism of ongoing AI applications in national defense. There is, however, a strong interest in this area by national defense sponsors of AI research.

5.5 Automatic Theorem-Proving

An important research topic in AI is the development of systems capable of automatically generating proofs of proposed mathematical conjectures given a set of well-defined axioms. Theorem-proving systems have been built in a number of areas of mathematics including calculus (Martin and Fateman, 1971), set theory (Bledsoe, 1971), and topology (Bledsoe and Bruell, 1974).

Various approaches to automated theorem-proving exist. *Resolution refutation* techniques provide a general mechanism for proof by contradiction that is theoretically guaranteed to find a proof if one exists, but is often infeasible in practical applications because of the amount of search required. *Rule-based deduction techniques* apply more heuristic approaches and principles to theorem-proving. These techniques are generally feasible, but are not by any means guaranteed to find a solution.

While systems for generating proofs of mathematical conjectures may not be of immediate interest to national defense applications, many of the techniques that have resulted from this work have been found to be useful in other areas. Resolution refutation techniques have been found to be a useful approach to intelligent retrieval from data bases. Some automatic programming is based on theorem-proving techniques, as are some approaches to robotic problem-solving.

5.6 Summary

For the most part, the technologies described in this chapter are not sufficiently mature to support the development of practical application systems. However, the long-term prospects for applying these technologies are very promising. As AI researchers develop the capabilities for systems to adapt to a changing problem environment, keep track of their own assumptions, automatically generate software programs, etc., it is clear that these capabilities will be incorporated into many application systems.

6
Development Environments for AI Systems

With the explosion of serious interest in AI system development and applications, a comparable increase of options has appeared for hardware and software environments in which AI systems can be developed and delivered. Indeed, options are appearing so quickly that any attempt to document a comparative evaluation of these options would be useless. Such an evaluation would likely be invalid by the time it appeared in print. We can, however, characterize the option space, which is precisely what is done in Sections 6.1 and 6.2. Section 6.3 provides a discussion of AI on microcomputers. This somewhat specialized topic is included to provide the reader with an appreciation of the true requirements for baseline processing requirements in which AI system implementation and delivery can occur.

6.1 Language Options for AI Systems

Regarding programming languages for AI system development, three classes of options exist: (1) general AI languages, such as LISP and PROLOG; (2) specialized AI languages, such as the expert system development packages noted in Chapter 1; and (3) "conventional" programming languages, such as Pascal, C, Ada, FORTRAN, etc.

6.1.1 General AI Languages

Although various general-purpose AI languages have been developed, LISP and PROLOG are, by far, the most widely used. LISP, invented in 1968 by John McCarthy, has been the computer language of choice for most research and development in AI (Barr and Feigenbaum, 1982). This has been particularly true in the U.S. Several important ideas behind the design of LISP have made it a useful tool for AI applications. LISP is a language designed for symbolic computation with the *linked list* as the basic data structure. It has a control structure based on function composition and recursion. An essential feature of LISP is that it allows for the representation of "code as data". That is, it provides the ability to interpret a LISP program as data, or data as a program.

The essential core of LISP can be defined by about 10 primitive functions, and the rest of the language can be defined in terms of these primitive functions. A LISP interpreter can thus be written in LISP itself and serve to define the language. It should be noted, however, that most practical LISP implementations include many built-in functions. As a result, several dialects of LISP exist that have significantly different features.

The extendability of LISP, the elegance and simplicity of its representation and control structures, and the resulting powerful interactive debugging capabilities that are usually included in LISP implementations all make it an ideal language for rapidly testing ideas and developing prototype systems. The approach of using a flexible and extendable language for quick experimentation and cyclic development has been characteristic of AI research and system development. This development strategy is sometimes referred to as *iterative refinement*.

PROLOG is a programming language based on mathematical first order logic, and has been used in a number of AI research and application programs (Clocksin and Mellish, 1981; Clark and McCabe, 1982). It is very popular in the European AI community, and has been selected to play a major role in the development of the Japanese Fifth Generation Machine project (see Chapter 10).

PROLOG has built-in mechanisms for deductive inference and a uniform representation for the facts and rules that make up a knowledge base. New facts can be derived from existing facts and rules via the built-in backtracking and theorem-proving mechanisms. The extendability and flexibility of PROLOG also makes it a powerful tool for describing and developing solutions for programmers and system developers by using the approach of iterative refinement.

6.1.2 Specialized AI Languages

Another category of languages available for use in AI system development are the tools and languages specifically designed for certain classes of applications. Many of these systems obtain their specialization in part from the fact that they presume a specific approach to knowledge representation and utilization. Most of the expert system development tools noted in Chapter 1 assume a production rule representation of knowledge. On the other hand, languages and tools such as KRL, FRL and K:base assume a frame-based approach to knowledge representation.

All of these specialized languages are designed to help the process of knowledge-based system design and development. They generally provide built-in features and packages (e.g., rule editors) to aid system design and knowledge engineering. Also, they generally interface with other languages, and so can be used as subsets of more general systems.

6.1.3 Conventional Languages

Conventional general-purpose languages, such as Pascal, FORTRAN, Ada and C have been around for many years. Although a number of AI-based programs have been written in these languages, they are relatively few in comparison to the number written in AI languages. Part of the reason for the lack of use of these conventional "procedural" languages is that of historical precedent—the first AI programs were written in languages such as LISP. Since then, most AI programs and also work in AI programming environments has been done in general languages.

One "problem" of conventional languages is that over the years they have become more specialized for dealing with number-crunching and data base storage and retrieval, while most AI applications are oriented toward symbolic processing and recursive problem-solving. Conventional languages, however, tend to have faster throughput than comparable AI languages, and also have the benefit of many years of fine-tuning and efficient compiler writing. In addition, there are fewer hardware requirements and restrictions when using languages such as FORTRAN than with more specialized languages such as PROLOG or LISP. As a result, conventional languages are good for run-time applications, but are extremely weak in the development area. Many developers have adapted the strategy of performing research and exploratory development in an AI-oriented language, and then reimplementing the system in a conventional language. For some types of AI systems, where the technology is well understood (e.g., rule-based expert consulting systems), many developers have gone straight to conventional languages. Some examples of this are provided in Section 6.3.

6.2 Hardware Options for AI Systems

A variety of options exist for hardware environments in which AI systems may be developed or delivered. These options differ in a variety of dimensions, including their basic processing power, the degree to which they have specialized AI language processing capability, how advanced their graphic capabilities for system development are, etc. Provided below are some categories of options for AI hardware.

6.2.1 Multiuser General-Purpose Systems

Until the advent of LISP machines (discussed below), most AI research and development was performed in general "mainframe" environments (e.g., DEC 10) where the system hosted many simultaneous users performing a variety of different tasks. Since access to mainframe environments is usually readily available and inexpensive for single users, this is likely to remain a prominent option for AI system developers. However, two common weaknesses of this type of environment include a relatively slow response time for interpretive languages such as LISP when system use is heavy, and a limited number of graphics terminals, which implies development may have to occur on a standard text terminal. Both of these tend to slow up development time which may be problematic if a rapid prototyping approach to system development is used. However, if the AI system is to be embedded as part of a larger software system resident on this environment, then it is often desirable to stay within the same environment.

6.2.2 LISP Machines

As noted in the section on AI languages, LISP is the most commonly used language for AI system development in the United States. To facilitate more efficient processing of this language, specialized LISP processors were developed. LISP machines are special-purpose AI development computers that are built around a LISP processor. They are normally single-user machines with high resolution monitors and special input devices (e.g., mouse). In addition to rapid processing capabilities, these machines are characterized by having a host of software development tools to support AI system development, including extensive graphic tools for user interface development.

6.2.3 General Single-User Workstations

Powerful single-user workstations provide many of the same capabilities as LISP machines. These workstations often have the same hardware architecture as that of multiuser systems, but with a software and operating system environment oriented toward single users. As with LISP machines, these machines also typically include high resolution monitors and special input devices. Also, because many have available one or more LISP dialects, AI system development tools are also available for these workstations.

6.2.4 Microcomputers

Microcomputer configurations (e.g., IBM PCs) have been used suprisingly often in AI system development, especially for expert systems. Microcomputer-based AI is a separate topic discussed in Section 6.3.

6.2.5 Parallel Processing Machines

Parallel processing machines represent a new type of computer architecture that will allow many simultaneous (possibly interrelated) processes to occur concurrently. In principle, these machines will have many times the processing power of more conventional architectures and are likely to become a primary environment of delivering AI and other types of systems in the future. In particular, if some of the ambitious goals of the DARPA Strategic Computing Program are to be met (see Chapter 10), then computers with extensive parallelism will become a common delivery environment.

6.3 Baseline Requirements for AI System Development

As noted previously, AI system development "traditionally" is done in LISP or PROLOG on either a mainframe or in a specialized LISP or PROLOG processing environment. The popularity of these languages and machines in AI has led many developers and sponsors to assume that these environments are essential for implementing AI systems. In reality, quite the opposite is true. While these powerful development environments are very convenient they are often not necessary.

In this section, AI work on microcomputers is explored. The goal of examining this somewhat specialized topic is to provide the reader with a sense of *baseline* hardware and language requirements for implementing many of the AI technologies discussed in previous chapters.

6.3.1 Expert Systems

A number of different software systems have been designed for implementing expert systems on microprocessor-based computer systems.

Many are problem-independent tools for building expert systems. They utilize a variety of inference mechanisms and knowledge representation schemes. Many are currently being used for a wide variety of expert system applications and are undergoing enhancements or improvements for future releases. They run on a variety of microprocessor-based computer systems, including the more popular personal computers with minifloppy or Winchester disk storage capabilities. Many of these systems can run on as little as 64K bytes of memory with small knowledge bases, but with more memory, larger knowledge bases are possible.

Table 6-1 shows some of the microcomputer expert system develop-

System	Developer	Hardware	Required Memory
ACLS	Intelligent Terminals Ltd.	APPLE II+ IBM PC	64K
AL/X	Intelligent Terminals Ltd.	APPLE II+ LSI 11/02 IBM PC	64K
APES	Department of Computing Imperial College University of London	IBM PC (any machine with Micro- PROLOG)	128K
DELTA	General Electric Research and Development Center	PDP 11/23	N/A
ERS	PAR Technology Corporation	IBM PC/XT	256K
GEN-X	General Electric Research and Development Center	IBM PC LSI 11	256K
K:base*	Gold Hill Computers	IBM PC	512K
M.1*	Teknowledge Inc.	IBM PC	128K
The Personal Consultant*	Texas Instruments Inc.	TI Professional Computer	512K
RuleMaster*	Radian Corporation	IBM PC/XT	512K
TIMM*	General Research Corporation	IBM PC/XT IBM PC/AT	640K

Table 6-1. Some expert system development tools available for microcomputers.

*M.1 and T.1 are trademarks of Teknowledge Inc. K:Base is a trademark of Gold Hill Computers Inc. The Personal Consultant is a trademark of Texas Instruments Inc. RuleMaster is a trademark of Radian Corporation. TIMM is a trademark of General Research Corporation.

System	Applications
ACLS	Induction of rules for classifying chess and game positions.
AL/X	Oil production platform fault diagnosis.
APES	Determining eligibility for Social Security benefits; pipe corrosion diagnosis; suitability of sites for dam construction; care for terminally ill patients.
DELTA	Troubleshooting for diesel locomotives; plastic injection molding equipment; process controller designs.
ERS	Interpreting military sensor information; aircraft design for maintainability evaluations.
GEN-X	Troubleshooting for jet engines; industrial process control.
K:base	Financial analysis of interest rate swaps.
M.1	Various demonstration systems for the knowledge engineering tutorial T.1.
RuleMaster	Severe-storm forecasting; transformer fault detection.
TIMM	Financial analysis; personnel management; insurance underwriting for yachts; tuning the VAX VMS operating system.

Table 6-2. Some microcomputer-based expert system applications.

ment tools available at the time of this writing. Table 6-2 lists some of the expert system applications that have been pursued using these tools. Note that AL/X, an expert system development tool that provides an inference engine, rule description language and consultation system capability similar to the well-known PROSPECTOR system, was the first commerically available inference engine.

6.3.2 Natural Language Interfaces

As will be discussed in Chapter 7, natural language interfaces apply natural language processing techniques to provide users with flexible and easy-to-use information access and retrieval from one or more data bases. Although various NLI systems exist, they all have in common some type of ability to automatically create a data base access program to answer a variety of questions that can be input by the user in a natural language format. Figure 6-1 provides a short example of a session with the Advance Query Facility (Mah, 1980) which provides natural language access to several data bases on Soviet aircraft.

Compared to the extensive work in microcomputer expert systems, the number of microcomputer-based natural language interface systems is small. The reason for this is not, as might at first be imagined, that microcomputers do not have the computational power necessary for this type of natural language processing. Rather, it is that the utility of developing such capabilities has in the past appeared somewhat limited. Since the goal of an NLI is generally to provide easy access to one or more data bases, then implementing an NLI on a stand-alone microcomputer that does not contain any important data bases has little value. However, the increasing power of microcomputers and the corresponding increase in the use of microcomputers for serious decision support applications is likely to result in a proliferation of important microcomputer-resident data bases. This in turn may result in the development of a number of natural language systems for microcomputers.

6.3.3 Machine Learning

Even in the area of machine learning, there is serious on-going work using microcomputers. One system in particular, ACLS (Analog Concept Learning System), is a general system that allows users to generate classification rules from examples or cases of a rule application. These classification rules are essentially IF-THEN production rules that are often found in expert systems. ACLS provides a capability for describing the attributes of a classification problem and a set of examples that illustrate the application of an as-yet unknown classification rule, and from the examples generates a rule consistent with the examples. It therefore provides an effective tool of knowledge engineering. A marketed version of ACLS called Expert-Ease is available for microcomputers under the UCSD P-System Operating System, which is available for most microcomputers.

6.3.4 Summary of Baseline Requirements

Although the systems described represent only some of the work on microcomputer-based AI systems, it is clear from the variety of these systems that implementing AI capabilities—such as expert systems—on microcomputers is a serious option for users and developers

```
"A Q F DEMONSTRATION"
"   VERSION   08-06-00   "

GRY>TELL ME ABOUT AIRCRAFT.

THIS IS A SOVIET FIGHTER AIRCRAFT DATA BASE, DERIVED
FROM UNCLASSIFIED PUBLICATIONS. IT DESCRIBES CREWS,
FUSELAGES, WINGS, ENGINES, ARMAMENTS, AND PERFORMANCE.

QRY>LIST FIGHTERS ORDERED BY NATO NAME.

ROLE      /NATO/ NAME
_______________________________________

F/B       FARMER         MIG-19
F/B       FARMER-C       MIG-19SF
F/B       FARMER-D       MIG-19PF
F/B       FARMER-D       MIG-19PM
FGA/T     FISHBED        MIG-21
FGA/T     FISHBED-C      MIG-21F
FGA/T     FISHBED-D      MIG-21PF
FGA/T     FISHBED-F      MIG-21PFK
FGA/T     FISHBED-J      MIG-21PFMA
FGA/T     FISHBED-J      MIG-21MF
FGA/T     FISHBED-K      MIG-21MF
FGA/T     FISHBED-L      MIG-21MF
AWF       FISHPOT        SU-9
FS/A      FLOGGER        MIG-23
FS/A      FLOGGER-B      MIG-23B
FS/A      FLOGGER-C      MIG-23U
F/B       FRESCO         MIG-17

QRY>WHAT ARE THE DIMENSIONS OF THE FARMER?
--------------------------------FT--------------FT--------------FT
MIG-19    FARMER    LH = 48.83      Ht = 13.19      Ws = 29.54
```

Figure 6-1. Sample AQF session.

of AI technology. Often, this option is not feasible, or if feasible, is not preferred. However, it usually deserves serious consideration. A detailed discussion of AI on microcomputers and the conditions under which they might be preferred to more powerful hardware environments is found in Barth and Lehner (1986) and Lehner and Barth (1985).

6.4 Summary

Overall, it is clear that the developer of an AI system has many options for the hardware and software environment in which to develop the

```
QRY>HOW MUCH DOES IT WEIGH?

                        WEIGHT
------------------------ LB
MIG-19     FARMER       19180

QRY>PRINT THE ARMAMENT OF THE FARMER.

-----------------------------------------------------------------
MIG-19          FARMER          "CFON"          1 AA MISSILE ALKALI
                                                3    CANNON
                                "CFON"          4 AS ROCKET
                                                2    BOMB
                                                3    CANNON
                                "CFON"          1 AS ROCKET
                                                2    BOMB
                                                3    CANNON

QRY>HOW FAST CAN IT FLY?

                                     SPEED
----------------------------------MACH ----------- FT
MIG-19          FARMER          0.50             80000
                                0.60             70000
                                0.75             60000
                                0.97             50000
                                1.24             40000
                                1.32             30000
                                1.09             20000
                                0.84             10000
                                0.66                 0

QRY>THE FRESCO?

                                     SPEED
----------------------------------MACH ----------- FT
MIG-17          FRESCO          0-39             40000
                                0-48             30000
                                0-62             20000
                                0-80             10000
                                0-92                 0
```

Figure 6-1. Sample AQF session (continued).

AI system. There is no true "standard" environment for AI system development. Rather, the selection of an environment should be based on a serious evaluation of the specific requirements of the type of systems that will be implemented.

Part 2
Applications of Artificial Intelligence in National Defense

In Part I, alternative type of AI systems and their capabilities were examined. Part II provides a mapping of these types of systems into specific application areas in national defense. A variety of on-going applications and new initiatives in AI will be examined.

In Chapters 7 through 9, a number of case studies of AI applications for national defense are presented. The studies were selected with several criteria in mind. First, they focus primarily on operating prototype systems. This makes it possible to discriminate the status of applied AI as it stands today from the many promises of the future. In a field as dynamic as AI, it is often hard to discriminate the successes that have in fact been achieved from the (usually optimistic, and sometimes exuberant) expectations and promises of its proponents and contributors. By focusing on case studies this bias toward optimism can be somewhat offset. One result of the focus on operational prototypes is that these case studies will tend to reflect relatively limited systems that are small in comparison to those systems envisioned by the result of some recent DoD initiatives. Second, these studies have all been documented in open-source, published literature, thus making it possible for the reader to collect more information than the sketchy descriptions provided in this book. Third, within the above two constraints, these case studies were selected to be representative of a great many initiatives to apply AI to national defense. Please note that at no point is there a suggestion that the particular systems described in the studies represent the best available. Judgments of this type are *not* an element of this book.

Chapters 10 and 11 turn to the future of AI. Chapter 10 examines some of the major DoD initiatives in AI today. By examining where the DoD is investing its AI resources, it is possible to identify the direction of AI applications in the near future. Finally, Chapter 11 re-examines the opportunities for applying AI in our national defense, and the challenges that must be overcome before these opportunities can be fulfilled. In particular, challenges to applied AI that are unique to national defense applications are discussed. Personal judgments about AI's future and its likely successes and failures in the national defense arena are reserved for the end of this final chapter.

7
Artificial Intelligence and Military Intelligence

In the national defense arena, the term "intelligence" refers to the collection, correlation and analysis of information needed to support command decision-making. Intelligence *collection* is the process of tasking sensors and other collectors, and of recording the information they return. *Correlation* or *fusion* is the process of translating bits and pieces of collected information into a single integrated picture of the present status of an area of concern. In a conventional battle, for instance, the product of *tactical fusion* is a "snapshot" of the battlefield. Intelligence *analysis* involves the generation of higher level interpretations of the fused information. For instance, it is the responsibility of a *tactical intelligence analyst* to predict which course of action a battlefield adversary may pursue.

In this chapter, some prototype applications of AI technology to problems in military intelligence are explored. In particular, specific applications of expert systems, natural language processing and image understanding techniques are examined.

7.1 Expert Systems for Intelligence Production

When expert systems first emerged, the most common application domains were for problems that involved inference and hypothesis testing, as opposed to decision-making and option selection. Diagnostic medical systems such as MYCIN, geological analysis systems such as PROSPECTOR, and others all attempted to diagnose the true state of the world from a variety of imprecise indicators. MYCIN, for instance, examined test results, patient symptoms, patient histories, etc. to determine the type of bacterial infection afflicting a patient. As a result, much expert system technology is oriented toward this class of problems.

Once past the collection stage, almost all military intelligence problems involve an attempt to determine the true state of the world from a set of imprecise and possibly inconsistent indicators. Intelligence fusion can be characterized as a process of translating large numbers of imprecise collection reports into a set of well-defined entities in the domain of interest. Each entity (e.g., enemy tank division in area X) represents a hypothesis about the state of the world, and each collection report (e.g., possible siting of armored unit in area X) represents a possible indicator of the truth of the hypothesis. Intelligence analysis involves translating correlated information into an evaluation of alternative hypotheses of present and future directions (e.g., "Is the enemy going to advance, defend or withdraw?").

Given the above characterization of military intelligence problems, it would appear that this class of problems is a natural application area for expert system technology. Following are some representative applications in tactical fusion and tactical intelligence analysis.

7.1.1 Two Expert Systems for Tactical Fusion

National defense establishments possess an incredible array of sensor and other devices for collecting data about the status and activity of a given entity on a battlefield. These devices collect many different types of data—normal or infrared photographs, radar images, reports of electronic emissions, acoustic information, etc. Each type of data may indicate something about the entity observed, but not everything that needs to be known. One device may indicate where something is, but not what it is. Another may be informative about what it is, but not where or how big it is (e.g., regiment or division). A third may be indicative of size (e.g., large armored unit), but not much about type (e.g., tank vs. motorized personnel carrier).

The problem of tactical fusion is to transform a continuous stream of literally thousands of reports into a list of military units and their locations that represents a single integrated picture of the battlefield. This is a massive data processing and interpretation problem that many believe cannot effectively be done manually. Expert systems is one class of advanced techniques that may support this process.

An Aid for Large Area Sensor Surveillance

The Automated Exploitation of Large Area Surveillance Sensor (AELASS) system is a production rule system developed by PAR Government Systems Corporation, for identifying the activities of military units based on surveillance data that is provided from a variety of collection devices. The description of this prototype system is drawn from Dyer, Frantz and Livingston (1983).

Problem domain

The increasing mobility of military ground forces makes essential the rapid collection and exploitation of sensor data to track enemy positions and to assess their intent. Advanced sensor systems exist to collect data sufficient to meet this requirement. However, because of the quantity of data and the short amount of time available to analyze it, computer support is required to correlate this data.

Many sensor systems possess two basic modes of operation: surveillance and tracking. The accurate tracking mode is used to determine a specific target location and identification. The surveillance mode provides general information on levels of activity over large areas. Most systems for automated tactical fusion rely on the accurate data derived from the tracking mode, but this limits the amount of terrain that can be covered. The AELASS system is oriented toward a better exploitation of surveillance mode data; this includes the detection of targets, maintenance of an activity level history, and estimates of enemy activity levels.

System description

AELASS identification of high-level military activity is performed in two parts. The first part of the system handles routine manipulations of data such as collection reports, and calculates averages of area activity levels over time. The output of this part of AELASS is the input to the second part of the system, which is the rule-based component. This second part of AELASS applies production rules to analyze measures of area movements to identify significant, high-level activities (e.g., motorized rifle division moving west). This second part is described below.

Figure 7-1 shows the architecture of the rule base component of AELASS. The rule base contains the knowledge that the system uses for its analysis of indicators and activities. It is implemented as a text file containing the rules which describe the conditions for identifying specific indicators and activities, and also the condition for setting flags used by AELASS event-processing algorithms. For a rule base with no internal errors, the parser will produce an inference net representation of the rules in the form of Pascal data structures that represent the relationships between the antecedent and consequent parts of the rules. These structures provide for a fast and efficient execution of rules, while the parser and rule base provide for easy modifications of the rules. The inference engine uses the inference net representation of the rule base to drive the analysis of high-level activities. It determines which rules are satisfied, and therefore which identifications or status flag changes should be performed as a result.

The rules in AELASS are based on military doctrine. They test alternative hypotheses about ground movements on the basis of their consistency with this doctrine. A simple notional example might be

```
IF
   high activity in rear area, and
   high activity newly detected, and
   other units not moving rearward
THEN
   assert new unit has entered rear area.
```

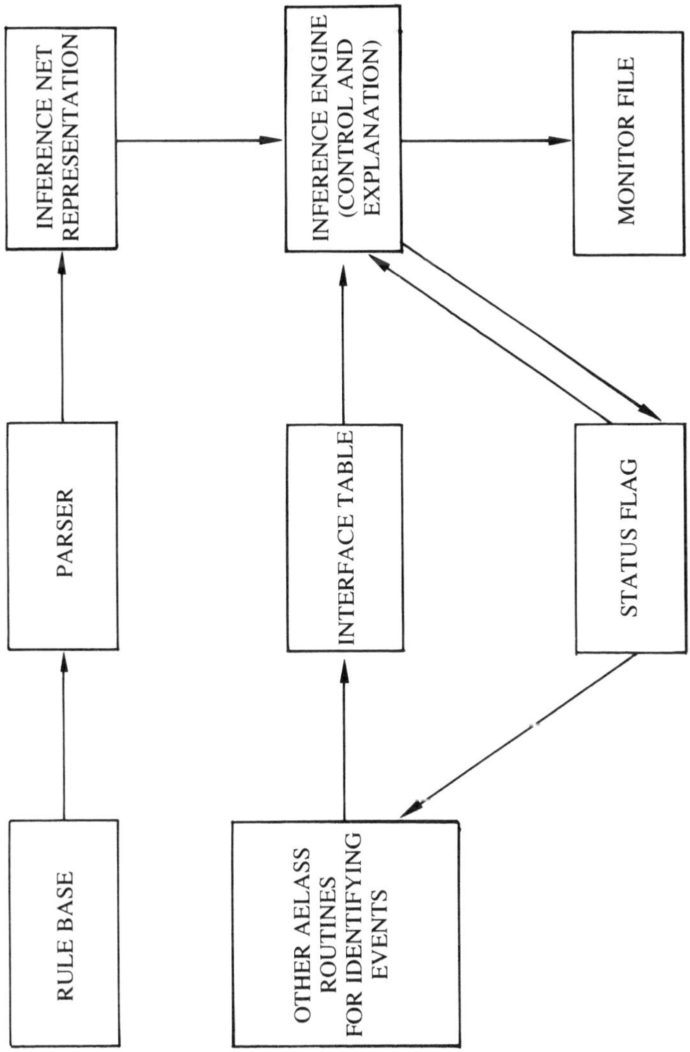

Figure 7-1. AELASS rule-based system component architecture.

This rule is consistent with a doctrine that does not use a selective withdrawal of units from the battle area.

The interface table and status flags provide the means of communication between the inference engine and other AELASS algorithms. The interface table is a list of counts and flags that are set and adjusted by the event-processing algorithms. For example, flags indicating the detection of certain types of emissions in various areas are stored in the interface table along the counts. Access to the interface table is depicted in the rule base as a function and its associated parameters are true.

Examples of status counts stored in the interface table include the number of active small and large units, the number of areas where units stop, the number of areas with high activity levels, etc. Fields that represent the detection of an event include identification of a garrison area, communication in the forward area, active radar in the forward area, deployment of artillery in the forward area, deployment of engineering units, deployment of ground forces, meteorological data, and large units in the rear area.

Status flags indicate one of four possible states of activity: normal (the default), movement of forces to contact, preparation for movement to contact, and commitment of forces. The consequent part of the rules may instruct the inference engine to change the settings of the status flag to be used as evidence, or antecedent conditions for the identification of some activity or formation of some intermediate hypothesis. The inference engine manages access to the interface table and status flags and changes the status flag setting when appropriate conditions in the rules are satisfied.

The output of the inference engine is a sequence of statements on the identifications or status changes identified by the rules. These statements are written to a monitor file which may be displayed to the user. In debug or explanation modes, the inference engine will also report the reason for each identification.

Comments on AELASS

AELASS is a tactical fusion application of the most commonly-used expert system techniques, namely, a goal-driven inference engine executing production rules. Although the rule base of the system as described in Dyer, et al. (1983) is small (around 50 rules), the general approach seems feasible for addressing this problem.

AELASS is also a good example of an *intelligent interface expert sys-*

tem (IES). As noted in Chapter 1, this class of expert systems is characterized by the fact that it automatically obtains its primitive input from an external baseline system(s) or data base(s), rather than by manual input. Furthermore, it functions as an "intelligent" processor of that information to support a user's ability to utilize the external data source. Put another way, AELASS is embedded within a larger automated tactical fusion environment, and its primary function is to provide its users with high-level analysis and an interface into that environment.

At present, we will not get into a discussion of the unique problems associated with IES system development. This is reserved for Chapter 11.

A Prototype System for Electronic Intelligence

Another example of a tactical fusion system is EXPRS (Expert Prolog System). The EXPRS work, as described in Pecora (1984), is oriented toward developing a general class of techniques that can be applied to the full spectrum problems in tactical fusion. This differs from the work described above in that AELASS was specifically designed for the problem addressed: exploitation of surveillance data.

EXPRS was developed at the Lockheed Palo Alto Research Center.

Problem domain

The general class of problems that the EXPRS work addresses is that of tactical fusion. The specific application area of the prototype system described by Pecora (1984) is that of combining multiple messages about possible ground units into a single picture. The EXPRS knowledge base is limited to messages derived from electronic intelligence (ELINT). ELINT messages describe detections of electronic emissions such as might be generated by a radar or radio transmitter.

System description

EXPRS combines the use of production rules for interpreting ELINT messages, with PROLOG clauses for describing perceived objects and their relationships. As noted in Chapters 1 and 6 respectively, production rules are a type of knowledge representation technique and PROLOG is an AI language based on predicate logic.

EXPRS uses an Attribute-Object-Value scheme for knowledge representation that permits for easy translation between production rules encoded in an English-like format and PROLOG clauses. The English-like rule descriptions use the following format.

```
rule 222:
IF
        (attribute1) of-the (object1) has-value (value1) and
        (attribute2) of-the (object2) has-value (value2) and
                                  .
                                  .
                                  .
THEN
     create-new (attribute) of-the (object) with-value (value) and
                         print (print-list)
```

EXPRS translates the preconditions of these rules into a set of PROLOG clauses of the form

```
attribute1(object1,?value)
```

and the create-new consequent into a clause of the form

```
asserta(attribute(object,value)).
```

Once this translation is made, EXPRS then maintains two isomorphic internal representations of the same knowledge: as formatted rules and as PROLOG clauses. The result of this is that a considerable amount of internal bookkeeping is handled automatically. Two important capabilities of EXPRS that are by-products of this automatic bookkeeping and the dual representation are described below.

Automatic description of objects

Whenever rule execution results in a new attribute, the function create-new automatically adds that new attribute to the list of valid attributes for that object. It also adds a clause of the form how-true(attribute(battery1,value)rule-number).

Whenever a user enters a command such as

DESCRIBE battery1,

EXPRS looks through the list of valid attributes for battery1. For each attribute it finds, it generates a PROLOG term of the form

attribute(battery1,?value),

and prints any clauses it finds that match this form. The how-true clauses are then searched to find a rule that leads to the assertion of an object value. A simple output might look as follows:

```
DESCRIBE emitter5
emitter5 description:
     belongs-to-battery             battery5     (rule 12)
     detection-time                 13:52
     emitter-designation            landroll
     radio-communication-node       node3        (rule 21).
```

Automatic generation of explanations

In any expert system, it is important to provide mechanisms for explaining to a user how the system obtained a particular result. Traditionally, this is done by identifying the specific rules that lead to the result. EXPRS handles rule tracing in two ways. First, whenever EXPRS generates a recommendation, it also prints out the rule that lead to the proposed recommendation. The user has an immediate option of asking, "Why rule x?" and will receive a printout of the rule that lead to the result. For instance, EXPRS might use rule 12 to determine that an emission report reflects a previously identified emitter. If a user enters

Why rule12,

the system would display rule information such as

```
     rule 12:
     IF
1)      designation of-the current-emitter has-value C-emitter, and
2)      position of-the-C-emitter has-value Position1, and
3)      position of-the Previous-emitter has-value Position2, and
4)      the-relation distance is-true-for [Position1, Position2, Distance], and
5)      Distance < 10.0, and
6)      belongs-to-battery of-the Previous-emitter has-value Battery1
    THEN
7)      create-new belongs-to-battery of-the C-emitter with-value Battery1.
```

This rule states that if a new emitter is within 10 kilometers of a previously reported emitter of the same type, then assert that it belongs to the old emitter.

If a user wants to know the specific values that lead to a result, the How command is used, for instance, using the clause numbers above,

How5

might lead to the display

Distance: 7.5 < 10,

indicating that the emitter was 7.5 kilometers from the previously identified emitter.

Comments on EXPRS

EXPRS is an attempt to build some general tools for handling problems in tactical fusion. As is evident in EXPRS, building tools often revolves around the ability to generate an appropriate approach to representing knowledge and facts in the problem domain of interest. Consequently, it is a good representative of how AI approaches problems such as tactical fusion.

It should be noted, however, that the EXPRS knowledge base is not as far developed as in some other AI systems that use ELINT processing. The system described in Barth, et al. (1984), for instance, has several hundred rules in its knowledge base and was evaluated empirically by operational users.

7.1.2 Intelligence Analysis

Intelligence analysis involves a higher level interpretation of correlated data to estimate intent and probable activities. In a tactical battlefield, for instance, intelligence analysis would be oriented toward such things as predicting enemy courses of action ("Are they going to attack in sector Y, and if so when?"). Performing this type of assessment is a complex problem that requires a broad knowledge base. The intelligence analyst performing this function is likely to use any of a variety of knowledge sources to make this type of assessment. These knowledge sources include:

- Knowledge of enemy tactics and the doctrinal employment of those tactics
- Knowledge of enemy resources, systems and capabilities and how they relate to the present environment (placement, mobility, status, etc.)
- Knowledge of friendly resources, systems and capabilities and how they relate to the present environment
- Knowledge of how the enemy may perceive friendly capabilities
- Knowledge of the broader political and economic environment that may impact or constrain enemy options (e.g., will not traverse neutral territory)
- Knowledge of an enemy commander's tendencies (e.g., conservative), etc.

As the above example illustrates, intelligence analysis is like many higher level analysis and decision problems in that the problem-solver must have a broad knowledge base to work from. This makes intelligence analysis a problematic application area for expert system technology. As noted in Chapter 1, expert system technology is generally limited to domains where there is a well-defined and bounded knowledge base. This suggests that the application of this technology must be limited to carefully selected subproblems characterized as being well-bounded and well-defined in their knowledge requirements. For instance, expert system techniques could effectively be used to assess the readiness and status of enemy units or perhaps to evaluate the impact of weather conditions on projected enemy movements.

Another approach to using expert systems in this domain is to apply this technology to enhancing other technologies—such as decision analysis—that may be more appropriate for high-level analysis and decision support. An aid that combines decision analysis and expert system techniques to support intelligence analysis is summarized below.

A Tactical Aid for Estimating Enemy Courses of Action

AI/ENCOA is a prototype decision aid, developed by PAR Government Systems Corporation, that is designed to assist Army tactical intelligence analysis in evaluating alternative Enemy Courses of Action (COAs). AI/ENCOA combines the use of additive multiattribute utility analysis (MAU) for course of action evaluations with rule-based procedures for assigning parameter values (scores and weights) to the MAU model. The description of this system is taken from Lehner, et al. (1984 and 1985).

Problem domain

At the Army division level, one of the most important intelligence analyses to perform is that of predicting whether or not the enemy forces opposing that division are going to attack. AI/ENCOA attempts to assess, for a friendly division sector, whether the opposing forces are going to engage in a primary attack, secondary attack, hold in a defensive posture, or withdraw. Furthermore, once it is determined that the opposing forces are going to engage in some type of attack, AI/ENCOA applies a further analysis to determine likely primary and secondary avenues of approach (AOAs).

System description

Functionally, AI/ENCOA is composed of two parts: a generic software package that implements a combined AI/MAU architecture, and two rule bases for analyzing different kinds of enemy activity.

Conceptually, the AI/MAU software has three interacting components: (1) an MAU model and analysis capability; (2) a user interface system, called the Attribute Manager, that permits users to characterize the decision situation facing them; (3) a set of composition or production rules that translate the description of the situation into appropriate scores and weights in the MAU model.

An MAU model is a decision analysis model composed of a strictly hierarchical set of evaluation factors. MAU models select among alternative options by comparing each option with these evaluation factors. Each factor in the hierarchy is given an importance weight, and each option is scored relative to other options.

The role of the attribute manager is to query the user as to the nature of the decision situation. It asks the user a series of questions about the specific problem he is addressing. Each question corresponds to an attribute in a predefined attribute list. User answers to the questions set the value of each attribute in the general attribute list. Users also have the option to select and answer only those few questions addressing specific, minimal changes in repetitive decision situations, thereby permitting them to quickly modify the status of the attribute list.

The role of the parameter assignment rules is to translate the information about the decision situation—encoded in the attribute list—into scores and weights in the MAU model. This rule base will be decomposed into independent rules sets that correspond to the nodes in the MAU hierarchy. For each node in the hierarchy, there is a set of composition rules that determine the value of the parameters associated with that node. The preconditions in each rule correspond to one or more attributes in the attribute list. The action resulting from each rule is the assignment or functional adjustment of the parameter value of the associated node in the hierarchy.

Figure 7-2 shows a simple example of an attribute, four parameter assignment functions, and a terminal MAU factor drawn from the ENCOA base described in the next section. The attribute definition defines a multiple choice question that will be asked the user. Based on the user's response, the variable f_of_f will be assigned the value 1,2, or 3. The values of the variables f_of_f1 through f_of_f4 are a function of f_of_f such that for the terminal factor (Fields_of _Fire) in the MAU model, scores for the four options (primary _attack, secondary_attack, defend, and withdraw) will be equal to the values of f_of_f1 through f_of_f4 respectively. Figure 7-3 shows

ATTRIBUTE DEFINITION	PARAMETER ASSIGNMENT RULES	MAU FACTOR DEFINITIONS
	function f_of_f1 = if f_of_f = 1 then 10 else if f_of_f = 2 then 80 else if f_of_f = 3 then 100 else 0;	
multiple_choice question f_of_f -> 'Fields of Fire' 1: 'Greater than 3000 Meters' 2: 'Between 1500 and 3000 Meters' 3: 'Less than 1500 Meters';	function f_of_f2 = if f_of_f = 1 then 40 else if f_of_f = 2 then 90 else if f_of_f = 3 then 100 else 0; function f_of_f3 = if f_of_f = 1 then 100 else if f_of_f = 2 then 90 else if f_of_f = 3 then 30 else 0;	factor FIELDS_OF_FIRE weight f_of_f parent terrain_factors primary_attack f_of_f1 secondary_attack f_of_f2 defend f_of_f3 withdraw f_of_f4;
	function f_of_f4 = if f_of_f = 1 then 0 else if f_of_f = 2 then 50 else if f_of_f = 3 then 60 else 0;	

Figure 7-2. Sample elements of AI/ENCOA rule base.

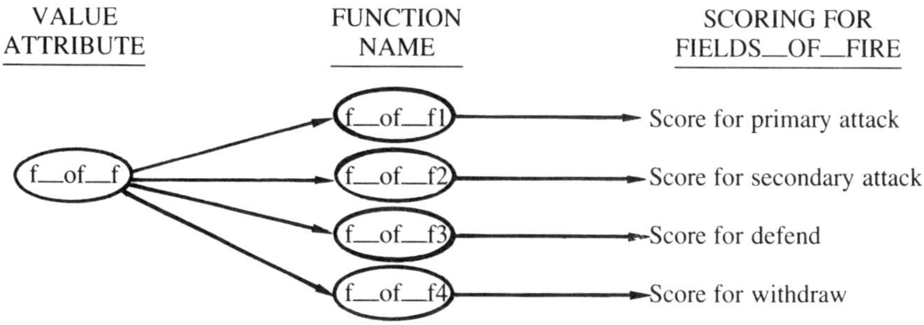

Figure 7-3. Example of mapping attributes to factor scores.

this functional relationship. Note also that the weight of the MAU factor Fields_of_Fire is equal to the value of wf_of_f which could be, in turn, a function of the answer to other questions.

This general approach to building an MAU-based expert system has two distinct advantages. First, the use of the Attribute Manager and parameter assignment rules make it possible to interface with users in terms and referencing with which they are familiar. In this regard, the user interface is very similar to that found in expert systems that do not contain a normative decision model. Second, as with other rule-based systems, this aid can be incrementally improved by simply adding, deleting, or modifying rules. This makes it possible to continually improve the aid's knowledge base, encoded as a combined normative MAU model and rule base.

AI/ENCOA presently has available two "rule bases" and models for evaluating possible enemy COAs. The first model addresses the question of whether the opposing forces facing a friendly division commander are likely to engage in a primary attack, secondary attack, remain in a defensive posture, or withdraw. This model is designed to help the intelligence analyst evaluate the severity of the threat that a friendly division commander may be facing. Figure 7-4 shows the MAU hierarchy corresponding to this first model. The second model is designed to help the analyst examine the support for a primary or secondary attack along each of the different enemy avenues of approach (AOAs) into a given friendly division sector once it has been determined there is a likelihood of an attack. That is, given that the analysis of the first model determines that a division commander is likely to be facing an attack, then the AOA model helps determine the degree of support

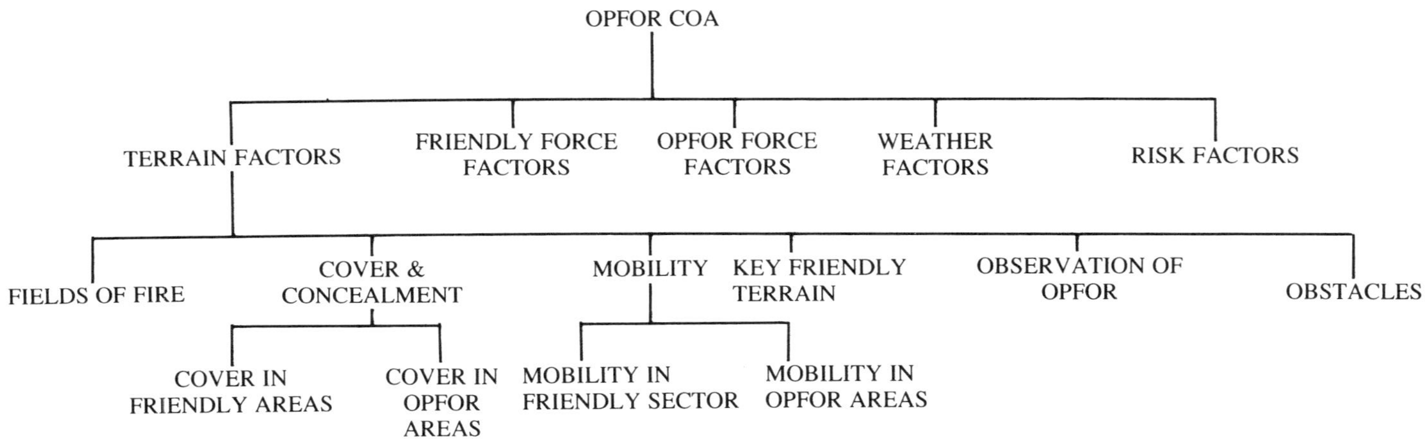

Figure 7-4. Portion of MAU structure for comparing primary attack, secondary attack, defense and withdraw options.

for a primary or secondary attack along each of the adversary's lines of advance. The MAU structure for the AOA evaluation model is similar to the one shown in Figure 7-4.

Figures 7-5 through 7-10 are annotated copies of displays from an interactive session with AI/ENCOA. This brief excerpt from a session with AI/ENCOA will give you a general feeling of how users interact with AI/ENCOA and will show you how the components of AI/ENCOA discussed in the previous two sections fit together. In paging through these figures, you will see a "single thread" example of managing attributes (Figures 7-5 and 7-6), assigning scores to options (Figure 7-7), and a textual result display of the Enemy COAs that received the strongest support (Figures 7-8 through 7-10). For each of these figures, the top half shows the AI/ENCOA display contents (actual displays are in color) and the bottom half a brief annotation. Each display contains several windows. The upper window is the primary output display. The bottom half provides windows that (1) describe the characteristics and status of the MAU factor presently being examined (The "Description" window), (2) show the scores for each option for the current MAU factors (the "ECOA Scores" window), and (3) show the menu options.

Comments on AI/ENCOA

AI/ENCOA provides an example of how expert system techniques can be combined with other technologies, in this case decision analysis, to support higher level intelligence functions. This aid differs from tactical fusion aids in that its goal is to assist and help organize a user's analytic processes, rather than to simply provide an automatically generated solution. This is because the aid revolves around a decision analytic model that provides a normative structure for performing COA analysis. The rule-based component of the system is there to enhance an analyst's ability to use the aid and the decision analytic model within it.

A general discussion of how expert system and decision analysis techniques can be combined in aid development is provided in Lehner, et al. (1985).

7.2 Natural Language Processing for Intelligence Data-Handling

Intelligence systems deal with enormous quantities of data, all of which must be transmitted to the proper recipients and recorded in the

```
Network Structure Diagram

                    ┌—TERRAIN FACTORS— — — — —┐
                    │                         │ ┌—FIELDS OF FIRE*
                    ├—FRIENDLY FORCE FACTORS  │ ├—COVER AND CONCEALMENT
                    │                         │ ├—MOBILITY
OPFOR COA— — — — — —┼—OPFOR FORCE FACTORS     └—┤
                    │                           ├—KEY FRIENDLY TERRAIN*
                    ├—WEATHER FACTORS           ├—OBSERVATION OF OPFOR*
                    └—RISK FACTORS              └—OBSTACLES*

              TERRAIN FACTORS — Subfactor of OPFOR COA

—Description— — — — — — — — — — ┤ ECOA Scores — — — — — — — — — — — — — — — ┤
  Wgt:  0.15    #Subfactors:  6 │ PRIMARY  SECONDARY  DEFEND  WITHDRAW │
 %Sol:     0       #Ans Ques:  0 │    0         0         0        0     │
  Sec:  sect 1  #Unans Ques:  8 │                                      │
—Solution Menu— — — — — — — — — ┴ — — — — — — — — — — — — — — — — — — ┘
       <Esc> Quit Menu        1. Answer Questions     4. Modify Sectors
         <?> HELP             2. Display Results      5. Save/Restore Answers
     <PrtSc> Print Screen     3. Analyze Results
```

In these excerpts, we will only consider the subproblem of evaluating the options of primary attack, secondary attack, defense or withdrawal from the perspective of TERRAIN FACTORS. The above display shows that we have moved down to the TERRAIN FACTORS part of the model.

Figure 7-5. First AI/ENCOA display.

```
Answering of Questions

Characterize Fields of Fire as:
      1 : Greater than 3000 Meters
      2 : Between 1500 and 3000 Meters
      3 : Less than 1500 Meters

2
-----------------------------------------------------------------------------
Characterize Cover and Concealment into Friendly Sector as:
      1 : Many (3 or More) Totally Covered and Concealed Routes
      2 : Few (1 or 2) Covered and Concealed Routes
      3 : Partially Covered and Concealed Routes
      4 : No Covered and Concealed Routes


                 TERRAIN FACTORS — Subfactor of OPFOR COA
—Description---------------------ECOA Scores------------------------------
  Wgt:  0.15    #Subfactors:  6 | PRIMARY  SECONDARY  DEFEND  WITHDRAW |
 %Sol:     0     #Ans Ques:   0 |    0         0         0        0    |
  Sec: sect 1  #Unans Ques:   8 |                                      |
—Question Menu:-------------------------------------------------------------
        <Esc> Quit Menu        1. Answer All Questions     4. Display Answers
          <?> HELP             2. Unanswered Questions     5. Erase Answers
      <PrtSc> Print Screen     3. Modify Answers
```

There are a total of eight questions relevant to evaluating TERRAIN FACTORS. This figure shows a display generated during the process of answering these questions. Note that the top question corresponds to the attribute shown in Figure 7-2.

Figure 7-6. Second AI/ENCOA display.

```
Display of Answers

Question Regarding:  Fields of Fire
    Current Status:  Between 1500 and 3000 Meters
--------------------------------------------------------------------
Question Regarding:  Cover and Concealment into Friendly Sector
    Current Status:  Few (1 or 2) Covered and Concealed Routes
--------------------------------------------------------------------
Question Regarding:  Cover and Concealment About OPFOR Assembly Areas
    Current Status:  No Covered Assembly Areas
--------------------------------------------------------------------
<More>

              TERRAIN FACTORS — Subfactor of OPFOR COA
—Description--------------ECOA Scores-------------------------------
 Wgt:  0.15   #Subfactors: 6 | PRIMARY  SECONDARY  DEFEND  WITHDRAW
%Sol:   100     #Ans Ques: 8 |    40        64       66       30
 Sec: sect 1  #Unans Ques: 0 |
—Question Menu:-----------------------------------------------------
      <Esc> Quit Menu        1. Answer All Questions    4. Display Answers
        <?> HELP             2. Unanswered Questions    5. Erase Answers
    <PrtSc> Print Screen     3. Modify Answers
```

The output window above shows the status of some of the attributes. Note that the Description window shows that all eight questions have been answered. Also, the ECOA Scores window shows the resulting score for each option on TERRAIN FACTORS.

Figure 7-7. Third AI/ENCOA display.

Summary of Scores: Numerical Display

Factor Name	Degree Solved	Rel. Weight	PRIMARY	ECOA Scores SECONDA	DEFEND	WITHDRA
Current Factor:						
FIELDS OF FIRE	100%	0.17	80	90	90	50

FIELDS OF FIRE — Bottom-Level Factor, Subfactor of TERRAIN FACTORS

—Description			ECOA Scores			
Wgt: 0.17	#Subfactors:	0	PRIMARY	SECONDARY	DEFEND	WITHDRAW
%Sol: 100	#Ans Ques:	1	80	90	90	50
Sec: sect 1	#Unans Ques:	0				

—Score Summarization Menu

<Esc> Quit Menu	1. Textual Summary	4. Cross Sector Summary
<?> HELP	2. Tabular Summary	
<PrtSc> Print Screen	3. Numerical Summary	

The above display shows the score assigned to each option for the terminal MAU factor FIELDS OF FIRE. These scores were assigned by the functions f_of_f1 through f_of_f4 shown in Figure 7-2.

Figure 7-8. Fourth AI/ENCOA display.

```
Summary of Scores: Textual Display

Given 100% of TERRAIN FACTORS factors are solved, best ECOA is DEFEND.
However, SECONDARY ATTACK is nearly as strong.
-----------------------------------------------------------------------
For this factor, DEFEND has strong support
with a score of 63.3.
The subfactors that provide strong relative support for DEFEND
over other ECOAs are FIELDS OF FIRE,
                     MOBILITY and
                     OBSTACLES.

The subfactors that provide support for other ECOAs over DEFEND
are COVER AND CONCEALMENT,
   KEY FRIENDLY TERRAIN and
   OBSERVATION OF OPFOR.
<More>

            TERRAIN FACTORS — Subfactor of OPFOR COA
|—Description--------------------ECOA Scores-----------------------------|
| Wgt:  0.15    #Subfactors:   6 | PRIMARY  SECONDARY  DEFEND  WITHDRAW   |
| %Sol:  100    #Ans Ques:     8 |   37        58        63       28      |
|  Sec: sect 1  #Unans Ques:   0 |                                        |
|—Score Summarization Menu ----------------------------------------------|
|       <Esc> Quit Menu      1. Textual Summary     4. Cross Sector Summary |
|         <?> HELP           2. Tabular Summary                            |
|     <PrtSc> Print Screen   3. Numerical Summary                          |
|-------------------------------------------------------------------------|
```

The above is a text description describing the results of AI/ENCOA's evaluation of the four options from the perspective of TERRAIN FACTORS. In this display, the factors that provide strong relative support for DEFEND are shown. Note that FIELDS OF FIRE is one of these factors.

Figure 7-9. Fifth AI/ENCOA display.

```
Summary of Scores: Textual Display

Given 100% of TERRAIN FACTORS factors are solved, best ECOA is DEFEND.
However, SECONDARY ATTACK is nearly as strong.
------------------------------------------------------------------------
For this factor, SECONDARY ATTACK has moderate support with a score of 58.3.
The subfactors that provide strong relative support for SECONDARY ATTACK
over other ECOAs are FIELDS OF FIRE,
                     COVER AND CONCEALMENT and
                     KEY FRIENDLY TERRAIN.

The subfactors that provide support for other ECOAs over SECONDARY ATTACK
are MOBILITY,
    OBSERVATION OF OPFOR and
    OBSTACLES.
<More>

                TERRAIN FACTORS — Subfactor of OPFOR COA
              Bottom-Level Factor, Subfactor of TERRAIN FACTORS

┌Description──────────────────┬ECOA Scores──────────────────────────────┐
│ Wgt:  0.15   #Subfactors:  6 │PRIMARY  SECONDARY   DEFEND   WITHDRAW│
│%Sol:   100   #Ans Ques:    8 │  37        58         63        28   │
│ Sec: sect 1  #Unans Ques:  0 │                                       │
├Score Summarization Menu──────┴───────────────────────────────────────┤
│       <Esc> Quit Menu       1. Textual Summary     4. Cross Sector Summary │
│         <?> HELP            2. Tabular Summary                        │
│     <PrtSc> Print Screen    3. Numerical Summary                      │
└──────────────────────────────────────────────────────────────────────┘
```

The above is a text description of why SECONDARY ATTACK also has moderate support.

Figure 7-10. Sixth AI/ENCOA display.

proper data bases. Handling this quantity of data can be a significant problem in itself. Unfortunately, it is not uncommon for an intelligence system to collect and accurately interpret critical intelligence data, and then fail to communicate that intelligence to a decision-maker in time. The following two subsections provide examples of the application of natural language processing techniques to intelligence data-handling.

7.2.1 Natural Language Interfaces

As indicated in Chapter 2, one of the most common application areas of AI natural language processing techniques is in the development of easy-to-use interfaces to one or more data bases or data base systems. Within our national defense system and its component organizations, there are an enormous number of data bases that are routinely used. Furthermore, it is becoming increasingly important that the information in these various data bases be made readily accessible to a broader spectrum of geographically separated users. To achieve this level of integration, two primary tasks must be completed: networking the data base systems, and providing a single user interface to these multiple data bases.

Natural language interface (NLI) systems provide one approach to achieving the second goal. Various NLI systems for data base access have been developed (see Harris, 1977; Hendrix, et al., 1978; Kaplan, 1979; Templeton, 1979; Waltz, 1975), all of which have in common some type of ability to automatically create a data base access program to answer a variety of questions which can be input by the user in a natural language format. Generally, these systems operate by integrating through a sequence of the intermediate representations of the natural language query, resulting in the final iteration in a well-defined data retrieval command or program. Usually, at least three stages are involved. The first applies grammatical parsing rules to determine the function of each word in the input sentences, thereby creating an internal structure that describes the relationship between individual words (e.g., ". . . this adjective modifies that noun which is the subject of . . ."). The second stage then translates the structured query resulting from the first stage into an internal language description of a target data base access sequence that defines the specific steps required to search for and retrieve the requested data. The third stage then translates the target data base access sequence into the data retrieval procedures specific to each data base. Since most NLIs are designed to provide a user with access to multiple data bases, similar access sequences after the second stage may result in very different data retrieval procedures for different data bases.

An Example of a NLI System for Data Base Retrieval

One example (perhaps the classic example) of this type of system is LADDER (Language Access to Distributed Data with Error Recovery). LADDER is a single system designed to answer natural language questions about the content of several data bases containing naval command and control information. Given a query (e.g., "Where is the Kitty Hawk?"), LADDER will identify the data base(s) that contain the relevant data, plan a sequence of data base queries to the data base system(s) with the data, establish link(s) over ARPANET to the identified data base systems, execute the planned query to retrieve the desired data, recover from an error if necessary, and format and present the answer to the original question to the user. The LADDER system was developed at SRI International, and has been operational since 1976.

Problem domain

Domain of the data bases that LADDER accesses contains data primarily about U.S. Navy ships, plus some foreign and merchant ships. LADDER itself, however, is generic and can be adapted to interface to other data bases.

System description

LADDER uses a number of the natural language processing techniques noted in Chapter 2. We will not attempt to explain in detail how it processes natural language in this section. Rather, the goal here is to illustrate its basic capabilities along with examples of those capabilities in action. The examples are drawn from Sacerdoti and Sagalowicz (1980).

LADDER allows the user to identify the ship(s) of interest in any of a variety of ways, such as by name, class, type, classification, etc. Examples of legal queries exercising this capability would include:

WHERE IS THE CHARLES F. ADAMS?
LIST ALL ALLEN CLASS SUBMARINES.
WHERE ARE ALL INTELLIGENCE COLLECTORS?

Ships can also be defined by various attributes and combinations of attributes, such as

WHERE ARE ALL THE AMERICAN CRUISERS?
LIST ALL CARRIERS MORE THAN 1000 FEET LONG.
WHAT ARE THE CRUISERS IN THE MED THAT ARE NOT AT READINESS RATING C1?

Any attribute that can be used to define a set of ships can also be information retrieved about a ship.

WHAT IS THE RADIO CALL SIGN OF THE FOX?

Complex queries that combine the above capabilities are also possible.

WHAT STANDARD DISPLACEMENT IS EACH OILER WITHIN 400 NAUTICAL MILES OF GIBRALTAR?

It is also possible to apply "calculations" to retrieved data and to display the result of these calculations to the user. To do this, a programmer codes a general function for performing the desired calculation and then specifies how a user must refer to that function in order to invoke it. For instance, phrases such as, "HOW MANY MILES", "HOW LONG", "BEST ROUTE", etc. all correspond to coded functions for calculating distance, time, routes, etc. Examples of queries of this type include

HOW MANY MILES IS THE CONSTELLATION FROM HER NEXT PORT OF CALL?
WHAT IS THE BEST ROUTE FOR THE SUNFISH TO THE SKORY?

Similar to the ability to invoke calculations, the user can invoke display routines that show retrieved data graphically. Phrases such as, "SHOW A MAP OF", "HIGHLIGHT", "FILL", etc. correspond to programmed functions for performing the referenced action. Assuming the proper display device or terminal is available, commands of the following type can be processed by LADDER:

SHOW A MAP OF RADIUS 500 MILES AROUND THE FOX.
DISPLAY ALL SHIPS WITHIN 500 MILES AROUND THE FOX.
FILL LAND MASSES.
DISPLAY THE SHIP NAMES.
DELETE MERCHANT SHIPS.

Finally, users can also extend the vocabulary of LADDER by using LADDER. Statements such as

DEFINE "TIN CAN" LIKE "DESTROYER."

will let references to "TIN CAN" and "DESTROYER" be used interchangeably. Defining phrases with variables can also be done. For instance, the command

DEFINE "HOW FAR APART ARE ROTTERDAM AND OSLO" LIKE "WHAT IS THE DISTANCE FROM ROTTERDAM TO OLSO",

will be processed in a manner that will let a question such as

HOW FAR APART ARE THE FOX AND THE KENNEDY?

to be correctly processed.

A basic capability of all natural language interface systems is that of letting users enter ellipsis. An ellipsis means some words in a sentence are left out, yet the full meaning of the sentence can be understood from the context of the conversation. An example of a legal sequence in LADDER is

WHAT IS THE LENGTH OF THE FOX?
WHAT ABOUT DRAFT?
WHAT ABOUT THE SANTA INEZ?
WHERE IS SHE?

The first query will retrieve the length of the Fox, the second the draft of the Fox, the third the draft of the Santa Inez, and the fourth will give the location of the Santa Inez.

Comments on LADDER

Overall, the examples of LADDER given above are fairly typical of the capabilities of NLI systems. Systems such as these can, in principle, be applied to any of a variety of data base access problems. Many of these capabilities are in fact becoming standard elements in data query languages.

7.2.2 Message Processing

Effective communication within a C3I system is essential. With the increasing automation of C3I systems, there has been an exponential increase in the amount of information that could *potentially* be distributed through systems. We say "potentially" because there are many bottlenecks in communication systems that prevent the free flow of information. One of these bottlenecks is the fact that much communication is in the form of text messages which must be read, understood and acted upon (disseminated) by communications officers. No matter how efficient these officers are, this process is time-consuming. Consequently, in crisis situations where the volume of time-sensitive high priority messages can be overwhelming, effective communication may break down. If a backlog begins to emerge, the officer must either let message dissemination be generally delayed for all messages, or arbitrarily delay certain subsets of messages to allow others to get through.

Various techniques exist and are being developed to help alleviate this problem. These techniques are for the most part based on surface characteristics of the message (e.g., key words, special headers) and not on the basis of a deeper understanding of the message content. As a result, despite the fact that many of these techniques are quite ingenious (see, for example, D'Amore and Mah, 1985), they are inherently limited.

AI-based natural language processing techniques provide the opportunity to develop message processing support capabilities that are based on a machine's ability to understand message content. This support can come in either of two forms: interpreting a message when it's received or reforming messages before they are sent. In both cases, the system is based on a capability to quickly and automatically interpret generated text that is ill-formed, misspelled, ungrammatical, etc. A prototype system representative of this approach is summarized below.

Message Processing for Generating Well-Formed Messages

The Naval Onboard Message Analyzer and Disambiguator (NOMAD) is a computer system designed to understand Naval messages. It supports message processing by working the message sender to produce well-formed messages.

Problem domain

The problem domain of NOMAD is essentially Navy English, a specialized subset of English which includes relevant types of events (e.g., military operations, orders, maneuvers and battles), objects (e.g., ships and aircraft) and their characteristics (e.g., weapons available, speed and direction). The naval message in Figure 7-11, from Granger, et al. (1983), is representative.

System description

NOMAD interacts with a user to help him or her generate a well-formed message. The user is responsible for initially entering the mes-

NARR/S3 WAS KILO DELTA AT ONSTA HOWEVER INFLIGHT TROUBLE SHOOTING CORRECTED COMPUTER PROBLEM AT 1019Z. A DISTRIBUTIVE FIELD WAS LAID IN ASSIGNED AREA AND LOFAR CONTACT WAS GAINED ON BY MOST BUOY IN PATTERN. A DIFAR BUOY BEARING INDICATED TGT CUS OF 253T. LSFT 727 WAS VECTORED IN FOR DIP ON CONTACT AT 2035Z, WHICH WAS LATER EVALUATED AS NON SUB. ADDITIONAL LOFAR CONTACT ON NORTHERN BUOYS INDICATED TGT WAS TRACKING 020T WITH A SPEED OF 4.5 KTS CALCULATED FROM MOTORSLOT FREQ. LSFT EXECUTED AIRPLAN 44. HELOS HELD CONTACT FOR A PERIOD OF 15 MINUTES BEFORE GOING COLD. SLVG 702 INVESTIGATED LOFAR PROBABILITY IN SW PORTION OF OPAREA WITH NEGATIVE RESULTS. ENROUTE TO CV SLVG 702 BELIEVED HE HAD A VISUAL SUB BASED S ON WATER DISTURBANCE, WHITE WATER, PLUS SNORKEL EXHAUST STOVE. ACFT DROPPED A SMOKE AND A DIFAR BUT ADP HAD FAILED.

Figure 7-11. Sample Navy message.

sage, after which NOMAD interprets and translates it into a clear message without ambiguities.

The basic approach to natural language understanding in NOMAD is that of semantic analysis. As discussed in Chapter 2, in the semantic analysis approach, the parsing of input text is driven by word meaning rather than by a set of grammatical rules. Consequently, NOMAD is heavily dependent on having a good dictionary of word meanings and word relationships, where the procedures for processing words are embedded in the dictionary. Since naval messages are often ungrammatical, the semantics-based approach to text processing seems quite reasonable.

Provided in Figure 7-12 are some sample input messages from a hypothetical communications officer on the Midway and corresponding NOMAD outputs. Note that when the input is ambiguous, NOMAD will generate more than one interpretation.

Comments on NOMAD

NOMAD is representative of natural language systems for message processing. Systems such as this can be applied to a variety of message processing problems, and thereby enhance the message throughput capabilities of C3I systems.

input: Periscope sighted by Constellation abt 2000 yds off port qtr, and hs helo vectored to datum. Green flares sighted after visual on periscope. Hs gained active ctc and conducted two attacks.

output
first: The Constellation identified an enemy submarine that was at 225 degrees 2000 yards from its location. A helicopter squadron pursued the enemy submarine. The helicopter squadron identified some green flares. By using an active sonar, the helicopter squadron identified the enemy submarine, and they fired twice at the enemy submarine.
best: (same as above, except 3rd line, which reads:
The Constellation identified some green flares.)

input: Took under fire with missiles.
output: We launched some missiles.

input: Two destroyers were sighted.
output: We sighted 2 enemy destroyers.

input: Midway fired talos at fanning.
output: We launched a talos missile at the Fanning
[a fast-frigate1076].

Figure 7-12. Sample NOMAD interaction.

The primary difficulty with applying a NOMAD-like system for message processing is that of developing a complete and accurate dictionary. The process of generating these dictionaries can be very time-consuming and labor-intensive, rather like knowledge engineering with an expert system. In addition, in domains where messages may cover a broad variety of topics, developing such dictionaries may not be feasible. Granger, et al. developed a vocabulary extension system (VOX) to ease the process of generating and modifying these dictionaries.

7.3 Image Understanding for Intelligence Production

Simply stated, image understanding is to look at a photograph (or image) and recognize the objects therein. Image interpretation and object recognition problems are clearly a major activity in any intelligence system supporting national defense interests. Problems ranging from interpreting a reconnaissance photograph for map production to providing real time warnings of incoming hostile threats all may involve some form of image understanding activity. It is natural therefore that there is a great deal of interest in generating automatic image understanding capabilities to support these activities.

Since image understanding is a prevalent activity that cuts across many problem domains, much of the AI work in image processing has been oriented toward developing general image understanding systems that can be adapted or tailored to a variety of needs. These systems can be conceived of as image processing test beds, where new techniques or application algorithms can be developed and tested within the confines of a single hardware/software environment. A number of these test beds exist in both universities and private industry. The industry systems are of particular interest here since they are primarily oriented toward providing an environment where practical image understanding systems can be implemented.

7.3.1 A Test Bed for Generating Image-Processing Systems

One example of an image understanding system/test bed oriented toward practical system development is the Advanced Feature Extrac-

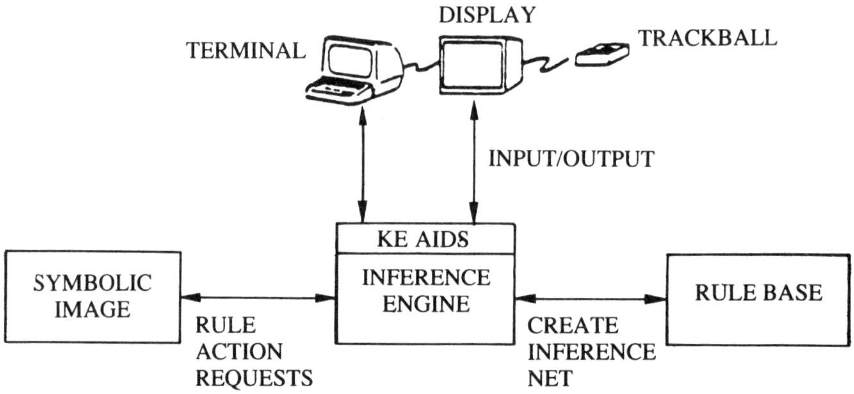

Figure 7-13. Symbolic image processing architecture in AFES.

tion System (AFES). As described in Lemmer and Mitchell (1984), a significant portion of AFES is oriented toward allowing a domain expert (e.g., photointerpreter) to develop a set of object recognition rules that AFES can use to interpret specific images. Specifically, AFES is oriented toward supporting the development of a computer-vision system with the architecture shown in Figure 7-13. This architecture contains three major components: a symbolic image, a rule base, and an inference engine.

The symbolic image is a semantic network characterization of a topographic area. In the semantic network, nodes correspond to regions, edges and attributes, while the links define relationships between nodes. For example, the simple network in Figure 7-14 indicates that Edge 1 is-part-of Region 1, and that Region 1 is-next-to Region 2.

Within AFES, the symbolic image is generated through the segmentation process. As discussed in Chapter 4, segmentation involves breaking an image into separate regions and edges that correspond to perceptually distinct areas within the image. Consequently, the symbolic image is a semantic network representation of the results of the segmentation process.

The rule base is a file containing a set of inference networks for testing possible interpretations of a region. The top node in each network corresponds to a possible interpretation of the object to which a specified region may correspond. Subnodes identify possible characteristics of the region that support the hypothesized interpretation. Figure

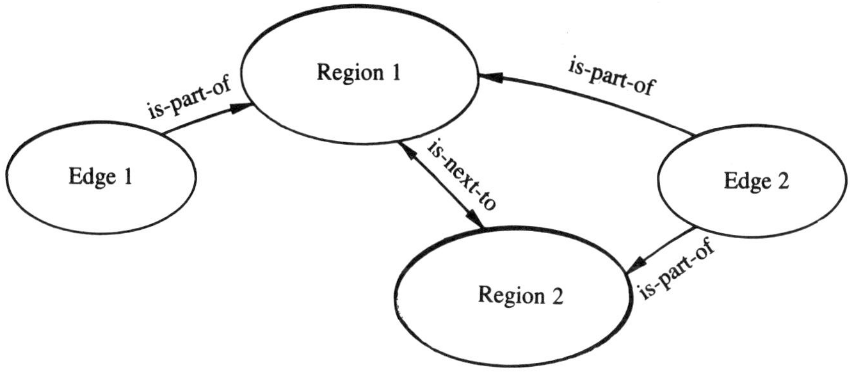

Figure 7-14. Elements of semantic network representation of segmented image.

7-15, for instance, provides part of a rule base for determining if a region corresponds to a storage tank.

Using AFES to develop a usable image understanding system for some domain of images is primarily a problem of building rule bases to recognize the objects of interest in that domain. As a result, AFES is embedded within several tools to support knowledge engineering. Specifically, these knowledge engineering tools help a photointerpreter build a rule base for detecting specified object types.

```
(hypothesis big_size
   text "current region has a large shape, test for correct area range"
   prior %0.1
   antecedents (and b_at_least b_less_than)
   consequences (oil_tank_size))

(evidence b_at_least
   text "current region is large enough to be a storage tank"
   prior %0.1
   consequents (big_size)
   comp-action (test
            ((current) ((area gt 190.0)))))

(evidence b_less_than
   text "current region is small enough to be a storage tank"
   prior %0.1
   consequents (big_size)
   comp-action (test
            ((current) ((area lt 360.0)))))
```

Figure 7-15. Elements of a rule base for image understanding.

The first of these tools provides users with feedback on the quality of a rule base. It does this by applying the rule base to a set of test case images where the objects and features have already been determined. Two forms of feedback are provided: (1) a list of the items in the image which had been misclassified, and (2) items in the image for which no clear classification was made. In either case, the imaging system developer is able to identify those objects, features, items, etc. that the existing rule base does not satisfactorily process.

The second tool interrogates an expert photointerpreter about how he or she detected a specific object within an image. Specifically, it inquiries as to the image characteristics that the expert felt were most relevant in identifying a specific object. The tool stores the results of these inquiries in the form of a series of "pseudo-rules" that list the object characteristics that are the preconditions in the heuristics that human experts use for item classification. The third tool assists users in translating these pseudo-rules into actual machine-usable rules (i.e., the relationships among the elements in the precondition lists are defined).

The fourth tool is graphic display-oriented. Specifically, it attempts to give the user an appreciation of the limits of a vision system's capabilities by showing a graphic representation of the information in the symbolic data base. Figure 4-1 is an example of this type of graphic display. By presenting images such as this, the rule base developer is able to determine the extent to which potentially diagnostic item characteristics will survive the image processing stage and be represented in the symbolic image.

Finally, the fifth tool simply provides an immediate test of individual new rules against a standard set of test cases. This is similar to the first tool, except that it focuses on individual rules rather than evaluating the entire rule set at once.

Overall, AFES as a whole and the AI components within it are generally representative of the test beds being developed for assisting in the creation of vision systems for specific domains.

7.4 Summary

The case studies in this chapter represent a small subset of the ongoing work in military intelligence applications of AI. These prototype applications address a broad variety of military intelligence problems and employ a broad spectrum of AI techniques.

These systems have several things in common. First, they are all prototype systems; few, if any, have matured to the stage of operational use. Second, a common problem faced by the developers of all of these systems is the need for extensive knowledge engineering. This includes not only expert systems, but also natural language systems, where extensive dictionaries and parsing rules must be defined, and image understanding systems, where the system needs a large number of image interpretation rules. Third, most of these systems are embedded; the data required by the expert system or image understanding system is obtained from the larger system.

8
Artificial Intelligence in Command and Control

Broadly stated, *command* is the determination of what to do, while *control* is the on-going management of the execution of a command. A military command and control (C^2) system is a decision-making network that reflects a hierarchical organization of C^2 nodes. Each C^2 node is responsible for the management of some portion of the available resources, where the higher level nodes are responsible for a correspondingly greater portion of the resources. Within a C^2 system, both command and control decision-making occurs at every level of the hierarchy. Command decisions at one level determine how to satisfy management decisions at a higher level. For instance, an Army Corps commander may determine that division X will engage in a frontal assault along corridor Y. This will result in various control decisions regarding the position and timing of the attack, along with the allocation of various resources to support the attack. Eventually, these various decisions are documented in orders to the division commander who must now make command decisions about how to engage in this frontal assault.

While command and control decision-making occurs at every level of a C^2 hierarchy, one factor that distinguishes these levels is time. As one moves from higher to lower levels of decision-making, both the time available for making decisions and the time horizon over which decisions apply decreases. In a conventional conflict, for instance, senior commanders will be primarily concerned with the determination of campaign plans with time horizons that may range from days to weeks. During the execution of a campaign, periodic (say, daily) orders will be transmitted to the next lower echelon, which in turn will periodically (say, every two hours) send new orders to its subordinate echelons, etc. until we get to C^2 nodes that actually execute the application of resources. These decision-makers often deal with the real-time decisions of an on-going battle.

In this chapter, the application of AI techniques to support C^2 decision-making is explored. The goal herein is to give the reader a general understanding of the types of AI applications that may be feasible for C^2 decision support. Several prototype AI-based decision aids are reviewed as case studies. The decisions these prototype systems support range from real-time weapon allocation decisions to the generation of long-term campaign plans. These case studies are provided to give the reader a practical sense of where AI can fit into the C^2 arena.

8.1 Expert Systems for Real-Time Decision Support

Loosely defined, expert systems are computer-based consultants that provide advice to users in complex real-world problem domains. One important application of this technology is in the development of "intelligent interface systems." This is any system that applies AI technology to assist a user in interfacing with one or more existing systems or data bases. As first noted in Chapter 1, an expert system that operates as an intelligent interface system can be described as a computer-based consultant that provides advice on real-world problems using information automatically drawn from the contents of an existing system(s) or data base(s). This is somewhat different from the more traditional applications of expert system technology which are not oriented toward intelligent interface problems. In these more traditional applications, the user must manually enter all problem-relevant data; otherwise, the expert system accesses a data base which exists only to serve the input requirements of the system.

8.1.1 The Real-Time Advisory System

The Real-Time Advisory System (RTAS) is a prototype expert system that operates as an intelligent interface system. It is designed to provide real-time tactical advice for Airborne AntiSubmarine Warfare (AASW) problems. In comparison to similar types of intelligent interface systems, such as the one envisioned to result from DARPA Pilot's Associate Program (see Chapter 10), the RTAS is very limited in scope. However, it is an existing system that is part of an operational training/simulation system called the Portable Aircrew Trainer (McWilliams and Ricard, 1986).

Problem Domain

The RTAS is designed to provide expert advice to the Tactical Coordinator (TACCO) position in AASW. The TACCO is responsible for locating, tracking, and neutralizing hostile naval units using the avionics available in P-3 and S-3 aircraft. The TACCO functions as both a command and control officer within this sphere of responsibility. While the TACCO monitors information provided by other aircrew members and other participating naval units, most information is provided using a single display screen called a Multipurpose Display (MPD). The MPD displays the relative locations of important objects (the aircraft, floating sonobuoys, suspected targets, etc.), data (aircraft and target velocities, distances, wind directions, instrument readouts, etc.), and other special conditions. Figure 8-1 shows the contents of the PAT (Portable Aircrew Trainer) simulation of the MPD for the P3-C aircraft.

System Description

The key components of the RTAS are shown in Figure 8-2. As with any expert system, the domain-specific problem-solving procedures are declaratively defined in a file called a *knowledge base*. As with many expert systems, the RTAS encodes knowledge in the form of a set of

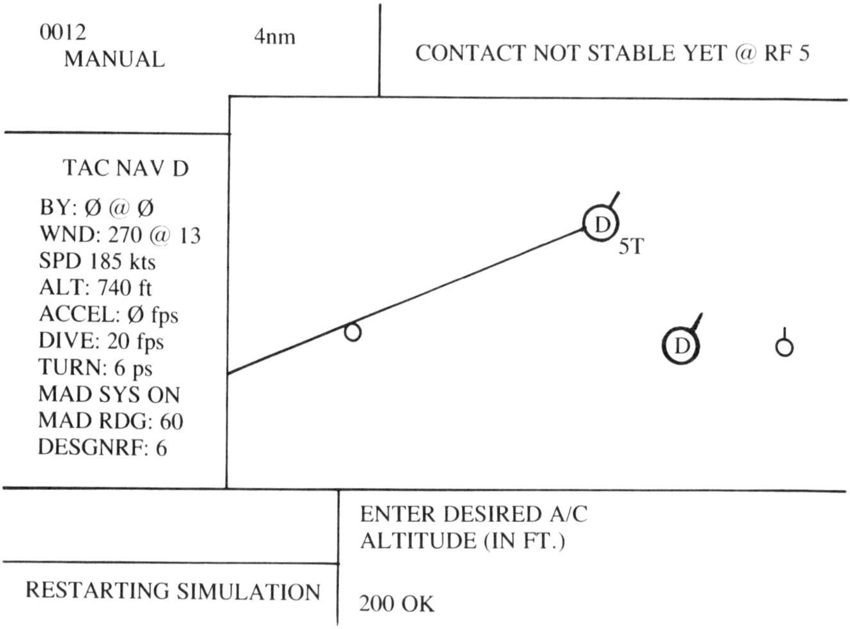

Figure 8-1. Example of a multipurpose display.

production rules. Consequently, the knowledge base is also referred to as a *rule base*. In the RTAS, the Rule Base (RB), the Rule Base Editor (RBE), and the Rule Base Compiler (RBC) provide the capability to encode and modify production rules expressed in the RTAS syntax. The rules declared in the RB relate the *evidence* (data concerning the state of the simulation as derived by the evidence manager) to the goals (hypotheses concerning what advice should be offered) in a precise way. The rules that define the relation between various pre- and post-conditions can be of a variety of forms (strictly Boolean, Fuzzy, Bayesian, or some other type of functional expression). The RBC converts the RTAS syntax to the proper internal form to be executed by the inference engine.

In any expert system, the role of an *inference engine* is to apply the problem-solving procedures encoded in the knowledge base to problem-specific data to generate problem-specific advice. In the RTAS, the Inference Engine (IE) is a rule-based deduction system that deduces pertinent advice to be given to the TACCO at any point in the simulated

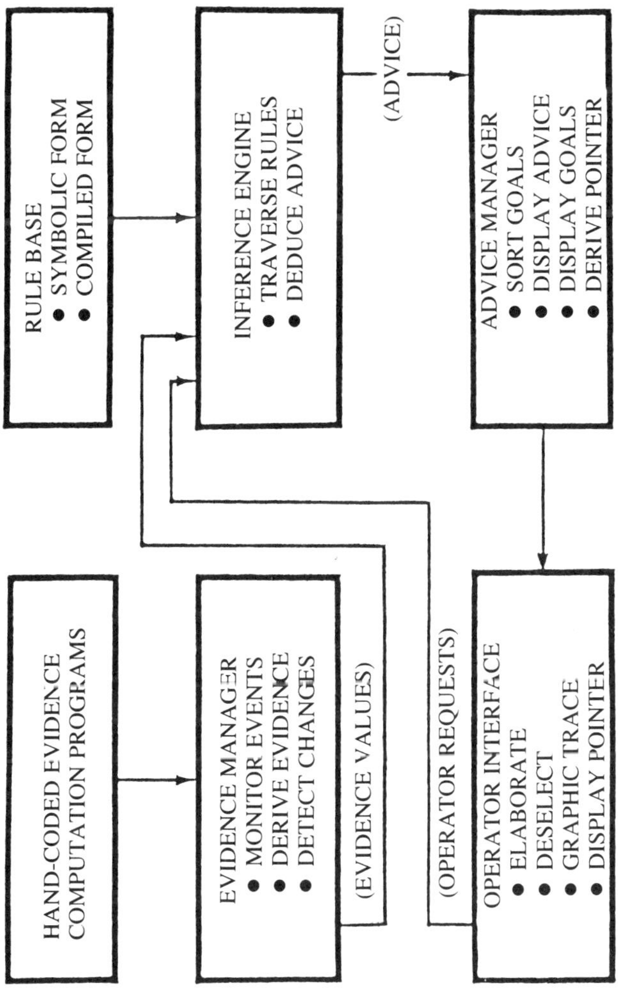

Figure 8-2. Components of ASW real-time advisory system.

scenario. The IE is invoked continuously or at the TACCO's request. Input in the IE consists of current values for each piece of evidence being monitored and past IE conclusions; output from the IE consists of deduced advice and the criticality of the advice.

The Evidence Manager (EM) in the RTAS computes the values for each evidence item being monitored. The computed values for all evidence items are ultimately combined by the IE to deduce the degree of urgency of each piece of advice the RTAS can give. Evidence item values are recomputed once each second based upon the values of the variables in the AASW simulations at that instant.

The Advice Manager (AM) and the Operator Interface (OI) work together to allow the TACCO and the IE to communicate. The AM sorts and passes advice to the OI; the OI provides advice, advice elaboration, and advice justification in various audio and visual forms. It also allows advice deselection—a graphic depiction of the goal tree—and a symbolic trace of the goals and evidence addressed in each single RTAS cycle.

System Operation

During an interactive session, the AASW simulator simulates the MPD available to the TACCO in the aircraft (see Figure 8-3). This simulated MPD is divided into several distinct areas. The square region in the center is a plot of the tactical situation (called the TACPLOT) as viewed from above. It shows the estimated relative location of the sonobuoys, the aircraft, and other significant objects. The border to the left of the screen displays dynamic information to the TACCO, including time, TACPLOT scale, status of autopilot, speed, altitude, and other flight information. The three rows across the top of the screen display information triggered by special events (called alerts) as they would be determined and reported by the aircraft's computers. The lower border of the TACPLOT allows the operator to respond to system requests for information.

These display elements are part of the standard P3-C MPD. Consequently, they reflect the data bases and existing sensor and tactical systems to which the TACCO must interface. The RTAS provides an expert interface to these data sources by calculating advice based on the data sources. In particular, in the RTAS-modified MPD, the most specific, current piece of advice is displayed in the lower-right corner of the screen. Updates of this advice are automatic once the RTAS system

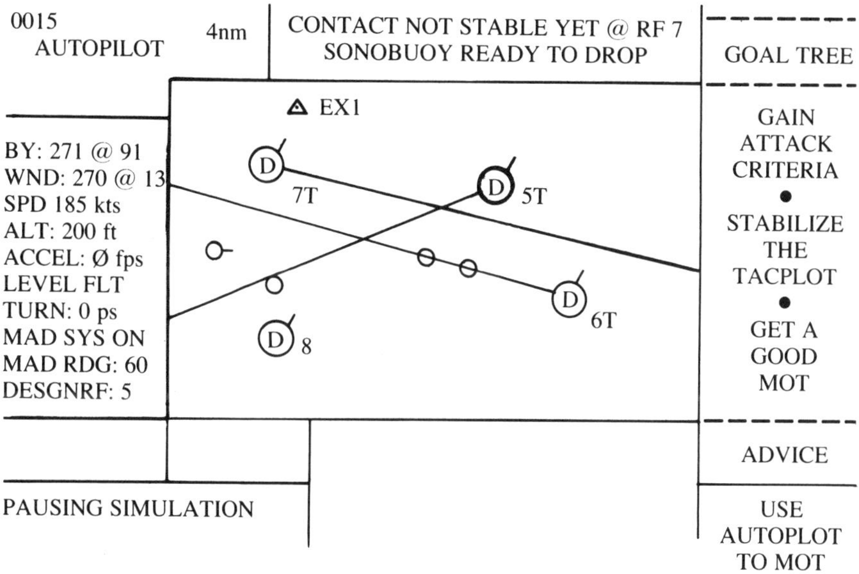

Figure 8-3. Goals and advice overlayed on a multipurpose display.

is engaged. Updates occur in real-time concurrent with changes in simulations of the tactical and sensor systems. A number of other features are provided for the operator, including the following:

Color-coded criticality

The RTAS advice is presented to the operator in a color appropriate to its current time criticality. Extremely critical advice—which should be responded to immediately—appears in red letters; moderately critical advice is in orange; and noncritical advice is in yellow.

Goal display

Upon request, the RTAS can use the right-hand edge of the TACPLOT to display the current hierarchy of goals served by the primary advice. Up to six high-level goals can be displayed in order in this window (see Figure 8-3).

Deselection

The operator can choose to ignore (pass) a particular goal or piece of advice by touching it on the display. The RTAS will recognize this as a deselection and will derive new advice and new goals not concerned with the deselected advice or goal. When this occurs, the word RESET appears at the bottom of the RTAS display to remind the operator that deselection is active. When RESET is touched, the advice status is restored.

Elaboration

Upon request, the operator can see an elaboration of the current advice. This requires the entire CRT screen, so it overlays the entire MPD. This elaboration feature displays the current goals and advice, the evidence used to determine a selected goal, and a brief text description of the purpose of a goal or the usefulness of a piece of evidence.

Symbolic trace

Upon request, a symbolic trace of the IE deductions can be viewed by the operator. The values of all evidence items and goals are displayed, as well as their deselection status.

Graphic rule base representation

This feature permits the operator to "step through" a traversal of the rule base and its current status by providing a graphic display of the paths and deductions. Operators have the option of displaying only those rules that lead to current advice, or examining any selected subset of the rule base.

Pointer display

RTAS also provides an array of pointers (small circle overlays on the TACPLOT) that draw the operator's attention to objects or locations on the TACPLOT that are referenced by the RTAS advice. Pointer functions that specify the TACPLOT overlays are defined as part of each rule in the rule base.

Comments on RTAS

Overall, RTAS is an excellent example of an expert system that operates as an intelligent interface, and demonstrates how expert system technology must be adapted to fulfill the requirements of an embedded system for providing useful advice in real-time. A more general discussion of the problems involved in developing embedded expert systems is found in Chapter 11.

8.2 Automated Planning for Resource Management

It is often stated that the key to military success is being able to bring the right forces to bear at the right time in the right place. To achieve this success, military decision-makers must have considerable flexibility to maneuver forces without a significant degradation of the capabilities of those forces. Consequently, a major aspect of C^2 decision-making is the scheduling and management of the massive quantities of resources required for effective military operations.

From the AI perspective, getting forces to the right place at the right time is a problem in *automated planning*, where planning is loosely defined as the ordering of actions over time. As discussed in Chapter 4, AI planners generate PERT-like structures called procedural networks that define the actions to be performed, order dependencies between those actions, and—if hierarchical planning is employed—provide the reasons for those actions. To generate a plan, these systems apply a knowledge base of goals and actions. The bases define the conditions under which goals and actions are feasible, and the constraints that need to be checked to test their feasibility. Consequently, a significant portion of planning involves the generation and testing of constraints. A successful plan is one that satisfies the defined plan objective without violating any constraints. Reasoning about expendable resources, for instance, can be handled by posting constraints on the minimum allowable level on that resource and constantly checking the plan against this constraint.

Chapter 4 points out some limitations of automated planners for problems involving the scheduling of action where there is no need to actively consider the actions of a competitive agent. Many C^2 resource management problems fall into this category. Two examples are overviewed below.

8.2.1 A System for Interactive Planning

SIPE (System for Interactive Planning and Execution Monitoring) is a general cooperative user/machine system for developing plans (Wilkins, 1984) developed at The Stanford Research Institute. Within SIPE, planning is treated as a problem in formulating procedural networks that accept all posted constraints. Parts of the procedural network can be entered manually (in which case SIPE checks for constraint violations) or automatically by SIPE.

Problem Domain

While SIPE is a generic system, one of the more complex domains it addresses is that of planning the operations involved in launching aircraft from an aircraft carrier. This involves such problems as determining the correct order for many—possibly parallel—actions, checking for air worthiness, and finding or generating a clear route to a fuel station and the runway.

System Description

To SIPE, a plan is a procedural network that defines an ordering of operations over time. Plans can be represented at several levels of detail. Figure 8-4 shows a simple plan at two different *levels of abstraction.*

Interactive plan generation with SIPE is a process of careful network construction. As noted above, nodes can be added to the procedural network by the user, in which case SIPE checks for consistency (explained below), or SIPE can automatically add nodes under the user's guidance. However, both SIPE and the user can only add nodes to the network that correspond to one of the legal operations defined in the SIPE knowledge base. Users cannot specify ad hoc operations (strip its parts and push it overboard).

SIPE assists users in constructing plans in several ways. These are overviewed below.

Constraint checking and resolution

Whenever a user proposes an action, SIPE checks it for consistency with the present state of the plan by determining if the prerequisite conditions for performing the action will be true. For instance, if the user proposes moving an aircraft to a refueling station, SIPE must determine if a clear path exists. This can be difficult since other, apparently unrelated, parts of the plan may involve repositioning several aircraft to clear other paths. An apparently clear path to the fuel station may unexpectedly be blocked just before it is needed. Since aircraft carriers are very crowded, and rapid response is essential in any emergency, this can be a significant problem. SIPE uses a general mechanism to reason

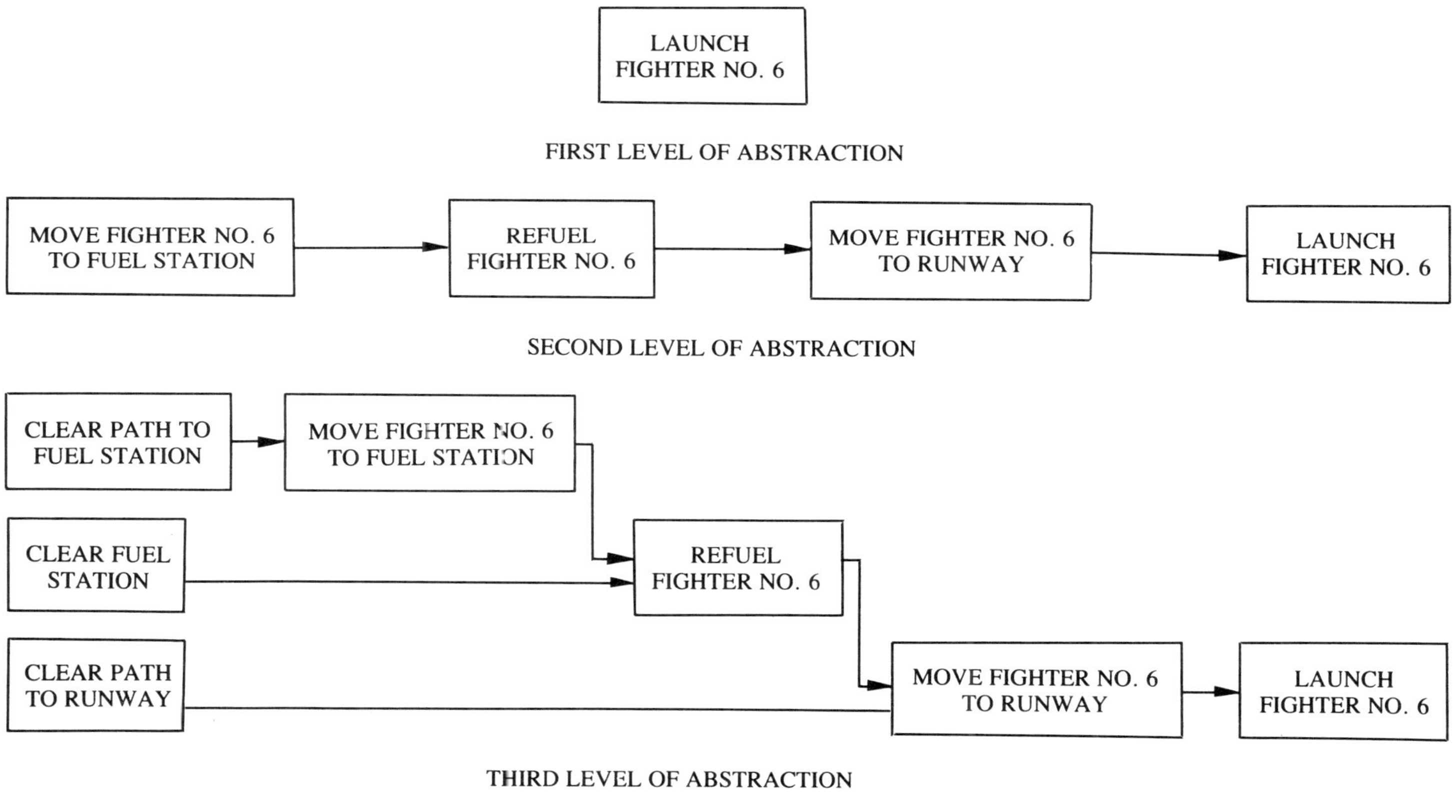

Figure 8-4. Notional plan shown at three levels of abstraction.

about conflicts in resource utilization, and a set of specialized deductive operators to reason about other types of possible inconsistencies.

Once inconsistencies are uncovered, SIPE can apply any of several mechanisms for resolving them. One approach is to interrelate the ordering of two parallel branches. If a precondition on one branch (clear path exists) is made false by the actions of another branch (other aircraft moved into clear path), then SIPE may recommend ordering the actions between these two independent branches (move aircraft through clear path before other aircraft is moved). To use this conflict resolution technique, SIPE must explicitly reason about the purpose of each action. Without recognizing that the purpose of a CLEAR_PATH operation is to allow the aircraft access to the fuel station, it could not recognize that the cleared path is no longer required after the aircraft is moved.

While the approach of resolving inconsistency by carefully ordering actions will always result in a feasible plan, it may also result in an inefficient plan. Imposing an ordering on previously independent actions results in an overall increase in the total time required to execute the plan. SIPE attempts to recognize conditions under which the ordering solution is not required. In particular, it attempts to determine if a discovered inconsistency is inherent to the evolving plan or to an inconsequential side-effect. In our example, for instance, SIPE may determine that blocking the clear path is a by-product of another clear path operation, and that the aircraft in question can simply be moved elsewhere.

In addition to finding harmful interactions between planning branches, SIPE also has a facility to sometimes detect and exploit helpful interactions. If the actions of one branch (e.g., clearing a path) also supports another branch (e.g., also clears a path for another aircraft), SIPE may recognize and attempt to exploit this interaction. In SIPE, exploiting an interaction is done by intertwining the order of actions among branches so that they will take advantage of the interaction. Note, however, that even "helpful" interactions may increase the total time required for plan execution, and so may not be exploited.

Automatic planning

SIPE can generate plans and subplans automatically. The interactive nature of SIPE comes from the user's ability to direct SIPE's automatic planner to solve specified subproblems. The user can, for instance, instruct SIPE to plan the detailed actions required to move a specified aircraft to a fueling station. SIPE will determine a set of consistent and sufficiently detailed actions to carry out the operation, and will display the results to the user in the form of a procedural subnetwork. Users can then modify this network to correct perceived difficulties. The user may, for instance, recognize that an as-yet unspecified part of the plan will move an aircraft through a path that the SIPE-generated plan has just blocked.

Plan explanation

Unlike most automated planning systems, SIPE is interactive. Consequently, SIPE also provides a capability to explain the rationale for each node in the plan. The mechanism for doing this is summarized below.

In the procedural network formalism, nodes represent actions to be performed or goals to be achieved. Whenever a node is planned to a greater level of detail, the expansion will generally consist of many subnodes. With each node, SIPE maintains a purpose attribute that specifies the important consequences that the goal or action identified in the node must achieve. For instance, CLEAR_A_PATH is a necessary subgoal of MOVE_TO_FUEL_STATION, while CLEAR_PATH(X) can be replaced with a similar goal to clear an entirely different path. By encoding the purpose of each goal or action in a plan, SIPE is able to explain to a user why a given part of the plan exists by displaying the purpose attribute of each of its constituent parts. Maintaining purpose information is also important for replanning; by understanding the purpose of each note, the consequences of removing a node are quickly deduced and the replanning requirements specified.

Alternative planning paths

In most domains, an objective can be achieved through any of a variety of action sequences. SIPE allows the generation of alternative plans by including "choice" nodes in the procedural networks. Consequently, SIPE users can simultaneously explore alternative paths for achieving the same goal without backtracking. A user can explore one path, and if the evolving plan begins to look inefficient, can quickly switch to another path.

Execution monitoring and replanning

During the execution of any operation, something unexpected always occurs. Consequently, a planner must monitor plan execution and perform replanning as necessary. SIPE addresses the issue of replanning.

During execution, SIPE allows a user or external computer system to specify executed actions and changes in the domain being modeled. (The term "modeled" is used to account for the fact that SIPE has no ability to trace conditions that are not in its dictionary of predicates, such as AIRCRAFT_ON_FIRE). At any point during plan execution, a condition may be true(false) that SIPE assumed would be false(true). When this occurs, SIPE examines all goals in the plan that are affected by this unexpected condition. Since SIPE records the purpose of each node in a plan, it can determine how the unexpected condition affects the plan, and can use its automatic planning routines to generate a new plan that maintains necessary goals and actions while replacing unnecessary nodes.

Comments on SIPE

SIPE is a direct extension of "mainstream" automatic action-planning techniques in AI for interactive planning. As such, it is a very representative example of how this class of AI techniques may be applied to C^2 decision problems. The approach appears applicable to any domain involving the scheduling of resources. Its value may be particularly appropriate for domains where the environment in very dynamic, such as in some battlefield applications. The ability to rapidly generate new or revised plans *that are internally consistent* would significantly enhance a commander's ability to respond to new situations and to work within an adversary's planning cycle.

A primary limitation of this approach is the requirement that the domain of application be completely modeled. SIPE is limited to operations defined within its knowledge base. To completely model the set of operators allowable to a logistics planner, for instance, would be a monumental task. The requirement for complete modeling of operators must either be relaxed or accepted. In the first case, the power of a system such as SIPE would be significantly reduced; constraint checking would be incomplete, and automatic planning could produce infeasible plans. On the other hand, complete modeling would be a significant investment. If this type of planning system is to be considered for potential application to a variety of C^2 decision domains, then the costs and benefits of using this technology need to be carefully evaluated.

8.2.2 A Resource Assignment System for Mission Planning

SIPE deals primarily with action sequencing, and only reasons about resources as necessary to determine the feasibility of an operation or to check for inconsistencies in a plan. In some military planning contexts, however, resource assignment is the predominant problem, while the ordering of actions is secondary in importance. One prototype system that focuses on resource assignment in a planning context is KNOBS (KNOwledge-Based System). KNOBS is an experimental system, developed by Mitre Corporation, that applies a variety of AI techniques to support the problem of planning tactical air missions (Engelman, Scarl and Berg, 1979). It is a fairly extensive system that incorporates action-planning techniques, production rule and frame-based formalisms for knowledge representation, and natural language processing. However, inasmuch as its primary objective is to support tactical mission planning, it appears appropriate to discuss it here.

Problem Domain

KNOBS addresses the problem of developing an air tasking order (ATO) for a tactical air force. An ATO is a daily order that assigns all the various units within a Tactical Air Force their missions for the day, along with a specific characterization of when and how each mission will be flown, the resources that will be used, coordination with other elements of the same mission, communication protocols, etc.

System Description

The basic product of KNOBS is a set of proposed mission taskings for the air resources in a Tactical Air Force. Complete mission taskings include the number and types of aircraft, the specific targets (primary and secondary), the originating air base, ordnance, time of departure, time over target, calling frequencies, etc. To support the development of feasible, consistent and coordinated mission taskings, KNOBS has a number of embedded capabilities. These are overviewed below.

Consistency checking

Figure 8-5 shows a single consistent interdiction mission. For a mission such as this to be feasible, it must be consistent with the expected world so that when executed, the aircraft and ordinance are available at the air base, the ordnance has the required probability of destruction

MSN NO:	MSN1220
TARGET:	BE145092-POL
PD:	.80
AIRCRAFT:	F-111E
ACNUMBER:	4
ORDNANCE:	B4
AOT:	1000
AIRBASE:	HAHN
STATUS:	UP
UNIT:	215TFW
TOT:	12-Jan-86 19:45:00
TDZ:	12-Jan-86 19:15:00
TRANSPONDER:	LINER6

Figure 8-5. Elements of a single-interdiction mission.

against the required target, the unit has had time to refurbish after a previous mission, etc.

In KNOBS, as in SIPE, consistency checking is done by posting and testing constraints. In KNOBS, most constraints are encoded as rules that identify conflicting situations. These constraint rules are prioritized so that major problems are noted and resolved before detailed planning problems are addressed. It will, for instance, recognize that a proposed air base does not have the necessary complement of the required aircraft type before examining ordnance availability. Since aircraft unavailability will require that the user select another air base, ordnance limitations of the initial air base need never be considered.

Explanation

KNOBS combines frame-based techniques for representing knowledge about objects (e.g., a particular type of missile battery) and relationship between objects (e.g., missile batteries of type X possess long track radar), with rules for identifying conflicts in the proposed utilization of resources. One result of this combined knowledge representation scheme is that KNOBS will sometimes proceed through a long sequence of reasoning steps before it identifies an inconsistency. KNOBS therefore includes a general interface that allows for the user-controlled interrogation of any part of the reasoning process. This differs from most expert systems that usually show a complete trace of the reasoning chain that leads to a conclusion.

Natural language data base queries

To facilitate a user's ability to query the data base of air resources tapped by KNOBS, a natural language interface facility is included. Although the implementation is somewhat different, the KNOBS interface has many of the same capabilities found in the LADDER system.

Option enumeration

During planning, a user may request advice on how to fulfill specified slots in a mission tasking. For instance, after selecting the type and number of aircraft, the user may want advice on which air base to select. To generate options, KNOBS proceeds through a process called enumeration. Enumeration involves reversing the constraint checking to identify all options that are consistent with the selections already made (e.g., only select air bases where the required aircraft are available). High priority constraints are used to filter out infeasible options, while low priority constraints are treated as preference rules that are used to rank-order the feasible options.

Rapid replanning

During plan execution, it often occurs that a proposed mission tasking becomes infeasible, such as when a friendly air base is attacking and is temporarily inactive. KNOBS can apply its consistency checking and enumeration capabilities to generate alternative options that satisfy the same mission profile (i.e., same aircraft and ordnance, but from a different air base).

Comments on KNOBS

The KNOBS system is an experimental prototype that explores the applicability of some AI techniques to mission planning. A second generation system called TEMPLAR (Tactical Expert Mission PLAnneR) is presently being developed by TRW that extends the KNOBS capability. TEMPLAR will be an advanced development model of an operational system that should be capable of handling full-scale, realistic mission tasking problems. The success of the TEMPLAR program will be a good indicator of the maturity of this technology, and the extent to which it can support a variety of operational C^2 planning problems.

8.3 Expert Systems for Battle Management

With the advent of the DARPA Strategic Computing Program (discussed in Chapter 10), battle management is rapidly becoming one of the most important applications of expert system technology. In this section, two prototype decision aids that support battle management functions are briefly overviewed. The first is a system to assist routine fire support planning functions. The second prototype is an attempt to provide rule-based support for the corps and division-level maneuver planning of ground forces, which is a high-level tactical decision problem.

8.3.1 A Fire Support Planning Aid

Battle (no apparent acronym) is a decision aid designed to assist fire support planning, and was developed by Slagle and Hamburger (1985) while at the Navy Center for Applied Research in Artificial Intelligence.

Problem Domain

Fire support is the application of resources such as artillery, surface-to-surface missiles, and close air support aircraft against enemy targets in support of ground force objectives. Fire support is a flexible resource that can be brought to bear against new objectives relatively fast. Consequently, fire support planning is a continuous process that results in frequent updates to the allocation and distribution of this resource. As noted in Slagle and Hamburger, the immediate objective of Battle is to improve the existing MIFASS (Marine Integrated Fire and Air Support System) intended to "provide for the establishment of fire and air support centers to plan, integrate, direct, and coordinate the fires of supporting arms."

System Description

Battle generates recommended fire support plans by employing two separate analytic phases. These two phases, along with a summary of their user interface characteristics, are provided below.

Phase I: weapon target effectiveness assessment

For a given fire support problem, Battle starts the analytic process by evaluating the effectiveness of each possible target/weapon pairing. It does this through the application of relevant "computation networks." A computation network is similar to an inference network with the exception that the possible set of mathematical expressions for calculating node values is expanded to include more than just probability/relative likelihood assessments. For instance, a simple CANREACH function can be included in the network to determine if a particular fire support weapon, in its present location, can in fact reach the target being considered. Figure 8-6 shows part of a sample computation network.

Battle applies computation networks, such as those in Figure 8-6, to each possible target/weapon pair to generate an assessment of the effectiveness of that weapon against the respective target. The results of the repeated application of these networks is then stored in a central data base of computational results. This data base provides the input to the Phase II analysis.

Phase II: generate composite fire support plans

Once the effectiveness of alternative target/weapon pairings is assessed, Battle then attempts to find an "optimum" distribution of resources against targets. Specifically, using user-defined target values, Battle attempts to find a distribution of fire support resources that will maximize destruction of the total target value.

Battle treats the optimization problem as a heuristic tree search problem, where each node in the tree represents a choice of applying a particular weapon to a particular target at a particular time. Since the number of possible decision points is prohibitively large for realistic problem domains, a pruning algorithm is embedded in the search mechanism. In AI, search pruning algorithms are often based on equations for estimating the best that a particular search branch will be able to achieve. In Battle, the pruning algorithm compares a proposed change to a distribution plan against other distribution plans that have already been developed. If the estimated best that the new emerging plan can do is not as good as that of previous plans, then that search branch is terminated.

One advantage of a tree search approach to optimization is that the search process often quickly leads to a reasonable "first pass" solution, but then continues to search for progressively better solutions. As a result, the search process can be tailored to the time limitations of individual decision problems. In addition, Battle users are permitted to constrain the solution generated in a variety of ways (e.g., ignore certain units, allocate or prevent allocation of a specific weapon against a spe-

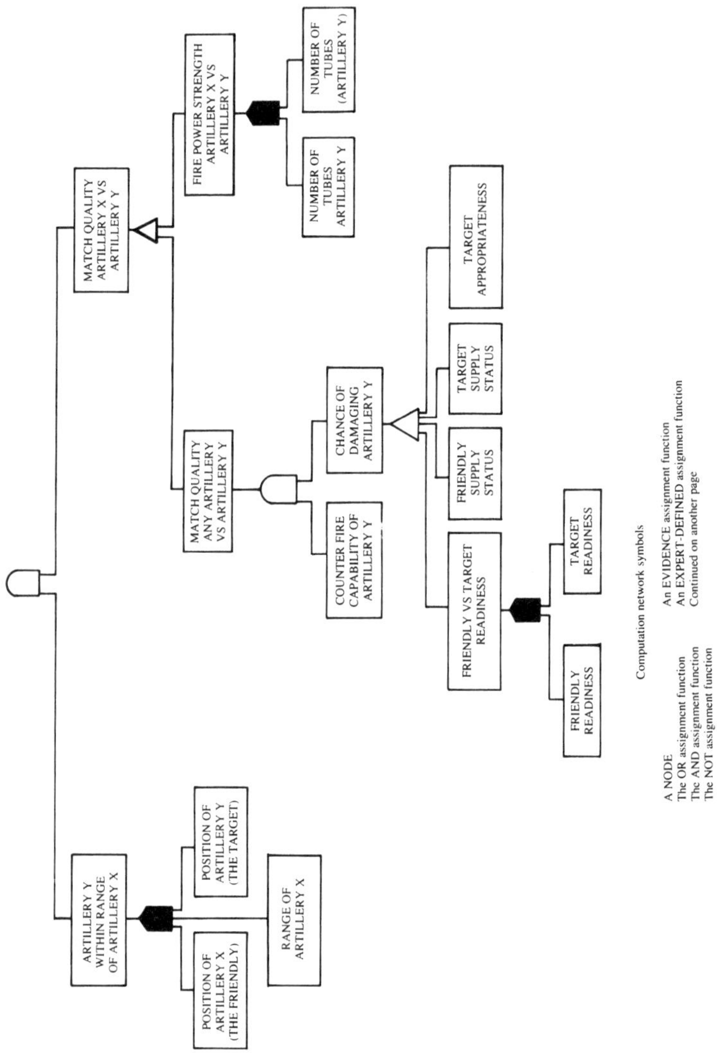

Figure 8-6. Portion of a computation network for effectiveness of artillery fire.

cific target, etc.). Imposing constraints on an optimization system employing tree search procedures, such as Battle, does not significantly reduce run-time efficiency. These constraints are easily incorporated into the pruning mechanism that rejects search branches inconsistent with the user-defined constraints.

Comments on Battle

Overall, Battle is a good example of the application of expert system technology to a battle management decision process. The focus of a system such as Battle is the repetitive application of a relatively small knowledge base (the computation networks) against a problem involving a large space of options. In problems such as these, "deep reasoning" with a great deal of knowledge is often not required. Rather, the rapid and repetitive application of "superficial reasoning" procedures is generally needed. Battle provided this type of decision support.

One aspect of Battle that bears comment is the reliance on simple algorithms and functions and the use of optimization procedures. Battle can arguably be perceived as an application of operations research (OR) techniques to fire support rather than of AI, or of both. Whether it's AI or not, Battle *does* reflect an increasing tendency in the decision aid development community to incorporate techniques from a variety of disciplines. As the supporters and developers of decision aids begin to recognize the complexity of decision support problems in areas such as C^2, they lose interest in maintaining pure technological boundaries. Anything that works is acceptable (see, for instance, Lehner, et al., 1985).

8.3.2 An Aid for Maneuver Planning

Within the AI-based decision aids arena, relatively few systems have been built that are designed to support senior command-level decision processes. This is in part because effective decision-making at senior levels usually involves a decision process that calls upon the decision-maker's broad experience and knowledge base. AI systems, on the other hand, are usually limited to areas where the knowledge required for problem-solving is both well-defined and limited in scope.

To the extent, therefore, that AI-based techniques can provide significant support to senior decision-makers, they must serve primarily in the role of local advisors or critics. That is, while an AI system may not be able to call upon a senior commander's extensive experience and knowledge base to generate decision options, it may be able to evaluate important options from the perspective of limited subproblems.

To build an aid of this type, one design problem that must be addressed is that of communication with the decision-maker. Since it is still the decision-maker's role to control option generation, the decision aid must have embedded within it a simple and natural mechanism for transcribing and representing user-generated options. This may be quite difficult for decision problems such as planning maneuver campaigns, where an option may be a complete concept of operations or a plan that involves many factors, sequencing considerations, levels of detail, etc.

One example of a system that adopts an "advice for limited subproblems" approach is TACPLAN (TACtical PLANner). TACPLAN is a prototype aid, under development, that is designed to support Army Corps staff personnel in the formulation of concepts of operations (see Andriole, 1984). It is comprised of three major conceptual components: (1) a videodisc-based user interface for map displays, (2) several analytical modules, and (3) a variety of knowledge bases of tactical planning and facts.

TACPLAN permits planners to enter their guidance and mission definition and then elicits judgments about aspects of the planning process (area characteristic assessments, relative combat power, etc.). TACPLAN checks these judgments against criteria defined in the rule bases. If the aid fires rules that lead to a disagreement with the planner, it will present the planner with the violated rule and offer advice (which may be rejected) to the planner about how to resolve the conflict.

The videodisc-based interface permits planners to annotate directly onto a map display by drawing candidate courses of action. TACPLAN then reads the annotation(s), accesses the relevant knowledge bases, and reports back to the planner about problems and opportunities presented by the candidate or hypothetical action(s).

8.4 Summary

The C^2 application systems described in this chapter share many of the same commonalities characteristic of the intelligence systems described in Chapter 7. First, virtually all C^2 systems are still in the prototype development stage. Few, if any, AI-based C^2 systems have demonstrated utility in a field environment. Second, because these systems are still in the prototype stage, developers have been working with relatively small knowledge bases. The extensive knowledge engineering necessary for a fieldworthy system must still be done. Third, most C^2 systems are embedded systems that obtain their problem-specific data from an external system or systems, and must therefore operate in coordination with the external system(s). As discussed in Chapter 11, there are a number of difficulties associated with building embedded AI systems that are not characteristic of the laboratory development of such systems.

The safest assessment one can make about the applicability in AI to command and control decision support is that the potential value is enormous, but has yet to be proven.

9
Artificial Intelligence in Autonomous Systems

An autonomous system is a self-directing/self-controlling system capable of perceiving, thinking, and of carrying out actions within its own environment. For instance, on a factory assembly line, a robot arm may be used to reject deformed products (e.g., a broken bottle). To perform this function, the autonomous system must "see" the product, "infer" that it is deformed, "plan" a set of actions to remove the deformed product, and then execute the planned actions. To achieve the ability to perform this function autonomously, the robot must have a rudimentary vision system, a pattern matcher for detecting specific shapes, and predefined action sequences for removing deformed products.

Within the national defense arena, autonomous systems that operate in controlled and limited environments—such as a factory—are of limited value. To support national defense interests, an autonomous system must be able to operate within the complex, real-world environment in which people operate. Within this broader context, interests range from the development of "smart" munitions that can find their targets despite obstacles and evasive tactics to autonomous ground vehicles that can independently maneuver across various types of terrain.

As autonomous systems move from limited and controlled environments to the outside world, their perceptual, inferential and planning capabilities must expand considerably. That is, the systems must exhibit intelligent adaptation to the world that humans inhabit. As a result, they need to exhibit a sophisticated level of artificial intelligence.

As indicated above, an autonomous system must have perceptual, inferential and planning capabilities to be able to operate within its environment. An *intelligent* autonomous system therefore presumably has AI subsystems for performing each of these functions. For example, it may have a computer vision/imaging system as the basis of its perceptual capability, an expert system for inferencing, and an automated planner to plan its movements. Intelligent autonomous systems therefore can be viewed in terms of an integration of AI and other technologies, rather than as a separate technological area in AI.

9.1 Autonomous Robot Vehicles

One example of an intelligent system is the autonomous robot vehicle presently being developed under DARPA sponsorship. Provided below is a discussion of the system envisioned by DARPA that illustrates the application of AI techniques within autonomous systems, and some of the technical problems that still remain. The programatics of this element of DARPA's strategic computing program in terms of its goals and anticipated schedule will be discussed later. The focus here remains technical.

The proposed Autonomous Land Vehicle (ALV) will be composed of three major parts: a vision system, a navigation system, and the vehicle platform itself (Kelly, 1984). As shown in Figure 9-1, the vision system is the primary source of information about the environment surrounding the ALV. A key element of this system will be a multispectral three-dimensional scanner. This scanner provides point (pixel) information that includes intensity and distance data, where the intensity data is specific to a spectral band. It is estimated that simultaneous scanning in

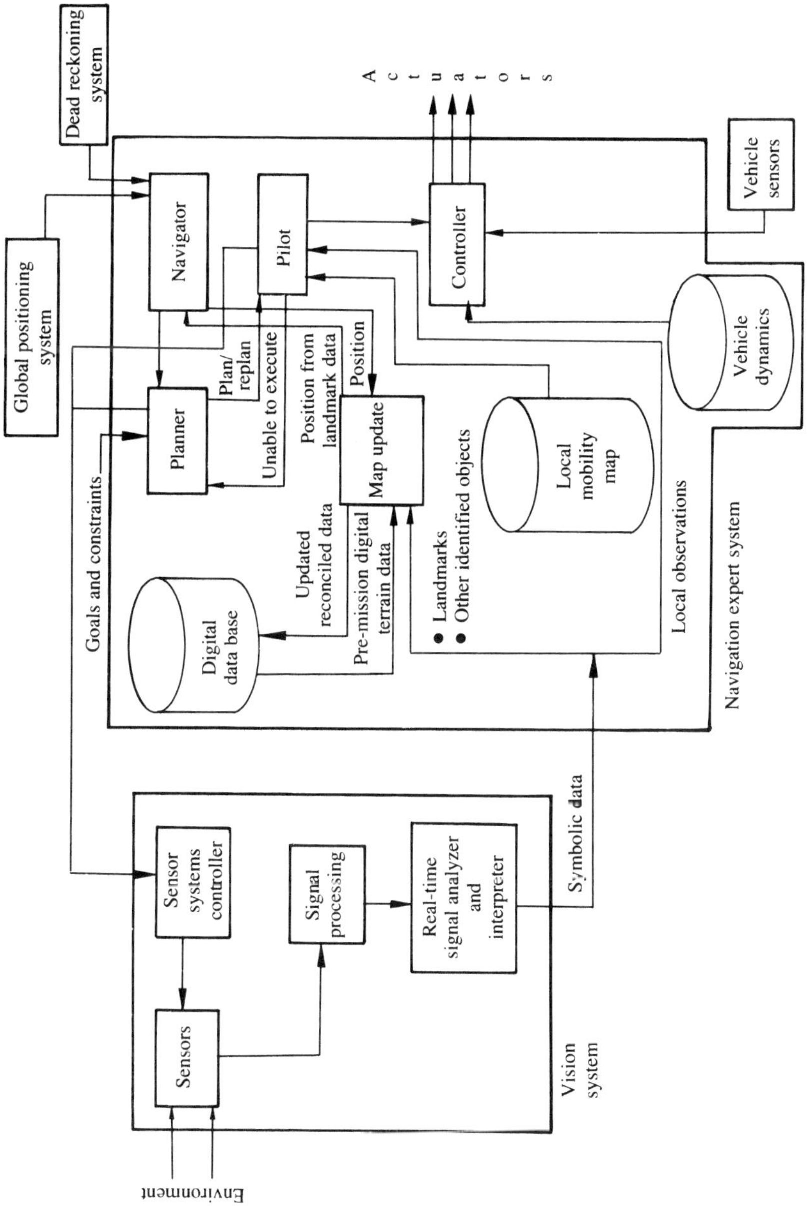

Figure 9-1. Conceptual components of an autonomous land vehicle.

about five spectral ranges will be needed, ranging from the visible to the infrared spectral ranges. Using an optical system with electronically addressible mirrors, the scanner will be able to quickly focus on selected areas. The desired response time for this adjustment is about one microsecond.

With the above scanning capability, the vision system will be able to collect a considerable amount of raw data about the world around it. The visual system must translate this data into an assessment of the location, size and type of objects in its range, and to produce a three-dimensional terrain profile. The estimated resolution required for the profile is about six inches.

To achieve this capability, the vision system first applies signal/image processing to detect various lines, edges and regions. For instance, applying image processing/segmentation procedures to the infrared image can generate thematic maps of the heat intensities of the various objects and regions near the ALV. As each image is processed, it is passed to the image understanding component of the system that must perform real-time analysis and interpretation of the area around the ALV. This component can use the thematic maps to identify regions corresponding to soil, roads, vegetation, etc. which may exhibit significant differences in thematic intensity.

The output of a vision system is a symbolic representation of the area in front of and near by the ALV. This constantly updated representation is then passed to the navigation system. Before discussing navigation, it is important to note that in the vision system, the sensors are controlled by the navigation system. The sensors will therefore be focused on any area of immediate interest to this system. This focusing significantly reduces the processing requirements of the vision system.

The navigation system is made up of components that all ultimately support route planning. The symbolic representation of local objects generated by the vision system is initially fed into a map-maker. The map-maker serves several functions. First, it uses and updates a digital terrain map composed initially of premission data. Second, it searches for predicted landmarks to provide feedback to the navigator on the position of the ALV. Third, it generates a local area mobility map used by the pilot to determine its precise local path. It is the pilot that attempts to execute the overall route plan generated by the planner. If local obstacles prohibit execution, the pilot provides this feedback and replanning is performed. Otherwise, instructions are sent to the controller that controls the actuators.

Overall, it is anticipated that the navigation system will be based primarily on expert system technology where the various functional components combined will have a large rule set, estimated by Kelly (1984) to be about 5,000 rules. To perform these functions in the short timespans necessary for ALV operations, an estimated improvement of computer execution time of about a factor of 100 will be required.

Regarding the vehicle itself, wheeled, tracked and legged vehicles are all being considered. The first vehicle being used is wheeled. It is expected, however, that a legged vehicle will also be attempted as a possible platform for the ALV.

Overall autonomous system development can be viewed as an application area of several technologies that must be merged into a single, integrated system. The ALV system described above is one proposed example of such an integration. It remains to be seen whether the AI technologies that comprise this system are sufficiently mature to support the development of a truly self-directing and autonomous robotic system.

9.2 AI and the Strategic Defense Initiative

In March of 1983, President Reagan proposed in a nationally televised speech the establishment of the Strategic Defense Initiative (SDI). The long-term objective of SDI is the design of a space-based system that can defend the United States against an intercontinental nuclear attack.

Since the initiation of SDI, a number of concepts have been proposed for its space-based architecture. These concepts all revolve around the existence of hundreds of Earth-orbiting satellites that possess various weapons and battle management capabilities. They all include a multilayer "defense-in-depth" conceptual architecture, where each layer of defense handles its own battle management and weapon allocation decisions. At a minimum, these concepts will assume boost, midcourse, and terminal layers of defense.

As with any national defense system, a space-based defense system must perform the full scope of C^2I (command, control and intelligence) functions. Intelligence functions range from the low-level detection and tracking of individual targets to an overall assessment of strategic situations. Command and control decisions range from an overall determination of strategy ("defend population centers") to the assignment of individual weapons to individual targets. A multilayer defense system requires that these functions be replicated for each defensive layer.

To the extent that a space-based system must incorporate traditional C^2I functions, it also provides a potential arena for applying many of the AI techniques discussed in previous chapters. Indeed, many of the specific system concepts discussed in the previous two sections should also apply to the SDI arena.

There are several features of the SDI program that make it a combinatorially more complex problem than any discussed in previous sections. First, the time duration for decision-making and action is very limited. The entire duration of a single wave of attack may be less than 30 minutes. Second, the space-based system will be faced with thousands of potential targets in this short duration. It must determine whether or not each potential target is in fact an active threat, plan a collective strategy against these threats, and perform all the necessary weapon allocation decisions to stop the attack. Third, the determination of weapon assignments must account for the fact that the set of available weapons is constantly changing as the satellites—which are not in geosynchronous orbit—move in and out of the conflict area.

Given the above characterization of a space-based defense system, it is clear that in order for such a system to effectively perform its function, it must be a largely autonomous system. As such, from the AI perspective, the SDI program represents an application domain that brings together a number of AI capabilities and techniques into a single autonomous system. A number of studies are under way to explore the potential of AI techniques to perform various C^2I functions.

In addition to performing various battle management and intelligence analysis functions, there are other areas where AI techniques can contribute to SDI. One involves the use of automatic programming techniques. It has been estimated that between 10 and 100 million lines of software code will need to be written to implement a space-based defense system. This is at least an order of magnitude greater than any single computer program/system in existence today. To develop the required software, therefore, it is necessary that a considerable amount of the code be generated through the use of automatic programming techniques, where the developer can enter a set of system specifications and the automatic programming system will generate code that performs according to those specifications.

A second important area is in the development of decision aids. As noted previously, the short time duration of a nuclear attack mandates that the defense system be largely autonomous. At the same time, however, politics and ethics are likely to require that the system and its various capabilities be under human control. This leaves us with the problem of keeping humans in the decision-making loop without slowing up the system to the point where it is ineffective. One approach is to provide human decision-makers with a variety of override capabilities where the system would perform its autonomous operation until otherwise instructed. To effectively perform this function, the humans will need considerable assistance in both understanding the evolving situation as well as in formulating an alternative response. AI-based decision aids could provide this assistance in that they could provide humans with an intelligent interface into the evolving situation and the defensive operations planned to deal with it.

Overall, SDI is an extremely ambitious project that provides many opportunities for the application of AI techniques to support our national defense interests. However, many have argued that the ability to develop such a system is well beyond the scope of current or near-future technology. So, it remains to be determined whether AI and other technologies are sufficient to meet the challenge of this initiative.

9.3 Summary

The development of intelligent autonomous systems is an important and active area of research in AI. It is also an application area of AI that is likely to reflect steady progress and some useful applications in the relatively near future. As noted above, AI-based autonomous systems must reflect a coordinated integration of several different types of AI systems, including expert systems, automated planning systems, and vision/image understanding systems. As research continues in these independent areas and new techniques are developed, these techniques will also find their way into autonomous systems.

In contrast to military intelligence and C^2 systems, there is comparatively little resistance to the area of automated systems performing some mundane functions (e.g., automated traversal of terrain). Consequently, as AI-based autonomous system research matures to the point of providing practical applications, there should be little to prevent the implementation and acceptance of such application systems. On the other hand, for large complex multicomponent systems, such as the largely-automated C^2I systems envisioned for SDI, there is still considerable question about the feasibility or acceptability of such systems.

10
National Defense Initiatives in Artificial Intelligence

The previous three chapters were oriented toward characterizing AI technology applications as they exist today. The goal was to give the reader an appreciation of what has been done with this technology by providing a broad set of representative examples. This is important, since the sheer number and breadth of AI programs in the national defense arena often makes it difficult to discriminate fact from promise. However, now that we have some of the facts, we can turn to the promises.

In this and the next chapter, we will look at where AI seems to be going in the next decade or so. To do this, we will examine some of the major initiatives involving AI and national defense. This is not purported to be a comprehensive survey. Rather, as in previous chapters, the goal is to provide the reader—through a set of representative examples—with a general sense of where things are going.

On-going AI initiatives will be explored from two perspectives. First, we will examine some of the major programs supported by DARPA and the three services. "Major" programs reflect, by definition, relatively significant investments; consequently, the objectives of these programs reflect overall DoD goals for the AI technologies and their application. Second, in Chapter 11, a broad survey of application areas will be provided. While some of these application areas are being pursued with a great deal more vigor than others, they represent the range of areas where AI techniques may be successfully applied in the near future.

10.1 DARPA Strategic Computing Program

Late in 1981, the Japanese announced a major project to develop fifth generation computer capability (JIPDEC, 1981). This fifth generation system is envisaged to be a series of interconnected data base and parallel processing machines accessed by means of an "intelligent interface machine" capable of accepting problem statements in either typed or spoken natural language (see Figure 10-1). The JIPDEC report that announced this program further states that:

> *The Fifth Generation Computer System will be knowledge information processing systems having problem-solving functions of a very high level. In these systems, intelligence will be greatly improved to approach that of a human being.*

This program is sponsoring a number of advanced concepts and projects with the goal of combining the various technological developments into a composite fifth generation machine by 1990.

The West responded to this announcement with both skepticism and consternation. Many Western scientists were skeptical about the feasibility of the program goals. At the same time, there was considerable consternation at the thought of possibly being by-passed in the development of advanced computing capabilities. Even assuming that the goals of the fifth generation program are far from satisfied, it was believed by some that the output of this program could still result in giving the Japanese a major technological lead on American companies (Bramer and Bramer, 1984; Feigenbaum and McCorduch, 1983).

As a result of Western concerns about losing ground in the devel-

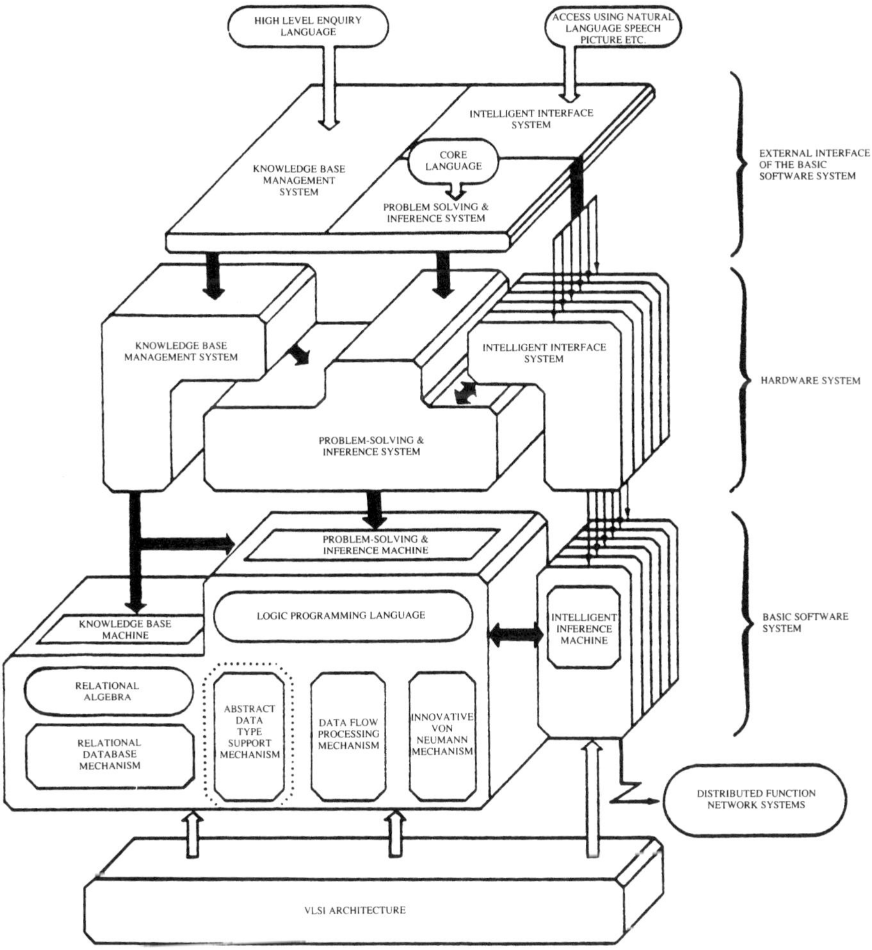

Figure 10-1. Basic configuration image of fifth generation computer systems.

opment of advanced computational capabilities, several projects were initiated. These include the Alvey Programme of the United Kingdom (Alvey Committee, 1982), the Western Europe ESPRIT effort, a cooperative consortium of several U.S. corporations called the Microelectronics and Computer Technology Corporation, and the DARPA Strategic Computing Program.

As described by Fields (1984), the major goal of the strategic computing program is to " . . . develop a broad base of machine intelligence technology to increase our national security." It is estimated that by 1994 DARPA will have expended more than a billion dollars toward achieving this goal.

The approach taken by DARPA is to push the development of practical systems and prototypes in three application areas: the development of autonomous systems that can traverse terrain without direct human assistance, the development of a cockpit system that can provide real-time advice to tactical fighter pilots, and the development of extensive battle management systems for both naval and ground forces. To meet the software and hardware processing requirements of these ambitious programs, advances in a number of key functional areas are required. These areas include vision, natural language processing, expert systems, automated planning and reasoning, and the integration of these technologies into automated navigational capabilities. Necessary improvements in these functional areas, in turn, require significant enhancements in the core hardware and software environments available for AI systems, which in turn require significant improvements in microelectronics. By focusing on a set of specific applications, DARPA has been able to select research and development goals to support that will provide for the advances in the technology base called for in the strategic computing program.

Provided below is a discussion of each of the application areas of the strategic computing program and a general discussion of the advanced hardware and software capabilities that are likely to result from this program.

10.1.1 Autonomous Land Vehicle

The overall objective of the Autonomous Robot Vehicle Program is to develop a self-piloting land vehicle that can autonomously cross a variety of terrains in pursuit of navigational objectives. While this may seem a grandiose concept, it appears fairly clear that many of the core technologies (vision, automatic navigation, simple automatic inferencing, utilization and generation of local cartographic information, etc.) for achieving this capability already exist in some form or another (see the discussion in Chapter 9). Therefore, the main thrust of this program is to push the development of these technologies to the level where they meet the performance requirements of an autonomous land vehicle (ALV) and to merge them into a single integrated system.

Figure 9-1 overviews the main conceptual components of the ALV, which will include a sophisticated vision system, expert navigation system, and a vehicle platform.

As discussed in Kelly (1984), this program will attempt to have a technological demonstration each year, beginning in 1985. These anticipated demonstrations have and will include the following:

- A 1985 demonstration of a wheeled vehicle that can follow an uncluttered road for a distance of 20 km and attain speeds up to 10 km/hr.
- A 1986 demonstration of road obstacle avoidance, with speeds of up to 20 km/hr on uncluttered sections of a road.
- A 1987 demonstration of cross-country route planning. The vehicle should be able to traverse 10 km of desert terrain with speeds of up to 5 km/hr.
- A 1988 demonstration of route planning over a road network where the vehicle uses visually detected landmarks to calculate its present position relative to the planned route. It should also be able to update its internal map enroute.
- A 1989 demonstration of cross-country traversal over a distance of 20 km, traveling at speeds of up to 10 km/hr. The vehicle should also be able to replan its route when confronted with major, impassable objects.
- A 1990 demonstration of combined road and open terrain traversal where the vehicle may travel up to 50 km and attain several navigational objectives.

As of the time of this writing, the 1985 and 1986 goals appear to have been met.

The above demonstration goals are for a wheeled vehicle. The ALV program is also exploring the possibility of using agile (i.e., legged) vehicles. This portion of the program seeks to have an agile vehicle capable of a walking speed of 5 mph, and a maximum "galloping" speed of 25 mph. The vehicle should be able to travel 100 miles without refueling and carry payloads of up to 500 lbs. It should be able to climb 60° slopes, cross 10-foot ditches, and climb 6-foot vertical steps.

10.1.2 Pilot's Associate Program

A second major application of DARPA's Strategic Computing Program is the development of a "Pilot's Associate," an intelligent, personalized, computer-based assistant to a pilot in such areas as the monitoring of system status, situation assessment, mission planning and tactics. The emphasis of this program is to build an associate that enhances mission effectiveness, in part by achieving increased lethality, survivability and safety while reducing the pilot's workload.

A conceptual architecture for the Pilot's Associate is shown in Figure 10-2. The initial construct shown in the figure has embedded within it four "expert systems". These are a situation assessment manager, tactical planning manager, mission planning manager, and systems status manager.

The function of the situation assessment manager is to provide an integrated assessment of the objects and other entities external to the aircraft that may have an impact on the on-going mission. It will deal primarily with the assessment and monitoring of the tactical environment by maintaining a current appraisal of enemy threats, targets, terrain and weather, and a continuous appraisal of the sufficiency of the own-aircraft status and available resources for meeting the mission demands.

The tactical planning manager provides advice on how to deal with the current tactical situation as perceived by the situation assessment manager. It will recommend a course of action composed of specific tactics, weapons choice, trajectory, and countermeasures to deal with current threats. In this manager, threats will be ranked according to their criticality of response. For each threat and proposed response, the manager will either wait for pilot confirmation or, in some conditions, automatically initiate a response with the opportunity for a pilot override. If sufficient time is available, it may provide the pilot with multiple recommended options from which the pilot can choose.

The mission planning manager will provide for the rapid adaption and revision of the current mission plan on the basis of the current on-board assessment and external command and intelligence inputs. It will compare the current situation and events and match them to a preplanned mission model. The impact of unexpected events, new commands and intelligence inputs are then translated into mission impacts which are processed to identify modifications to the original mission plan which will circumvent unexpected problems, yet still achieve the goals of the original mission.

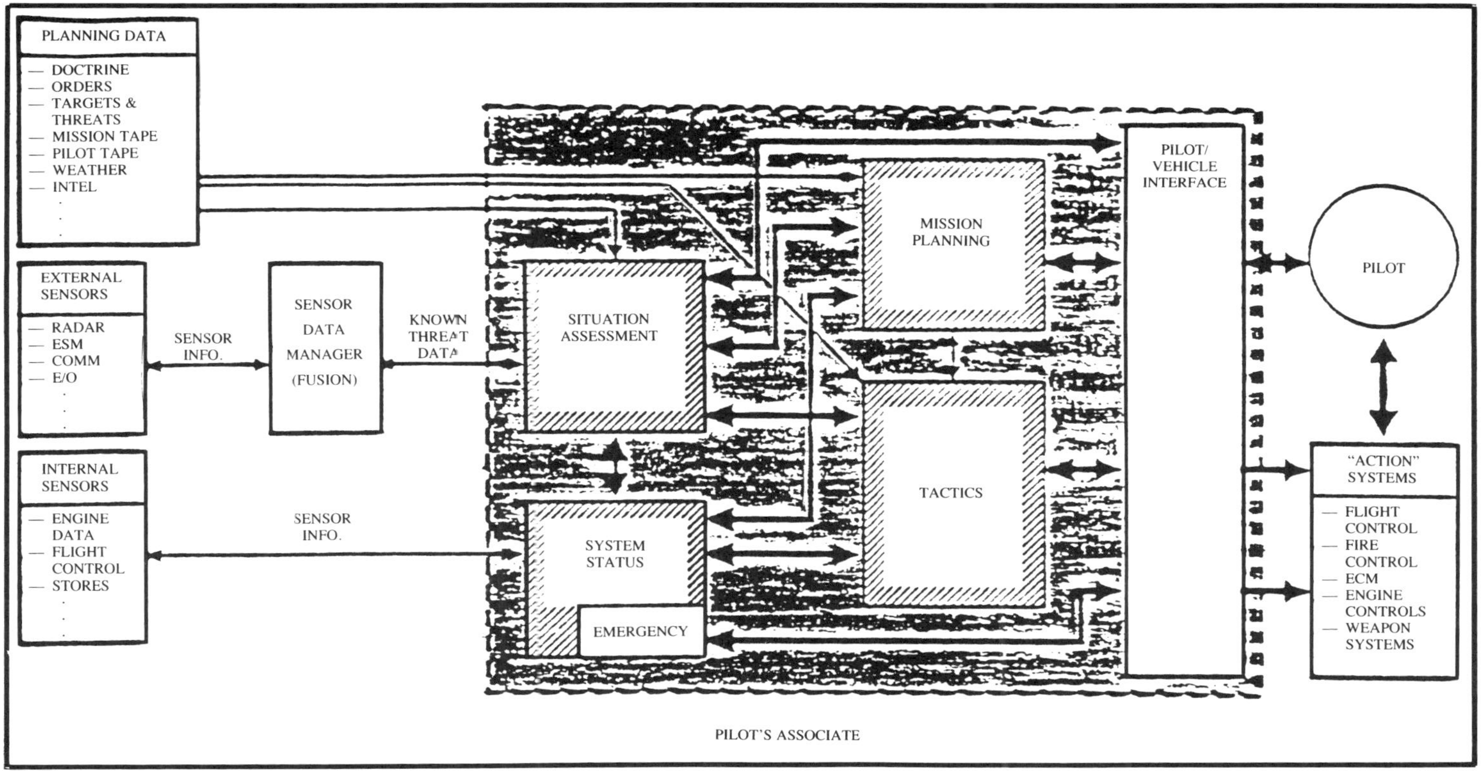
PLANNING DATA
— DOCTRINE
— ORDERS
— TARGETS & THREATS
— MISSION TAPE
— PILOT TAPE
— WEATHER
— INTEL
EXTERNAL SENSORS
— RADAR
— ESM
— COMM
— E/O
INTERNAL SENSORS
— ENGINE DATA
— FLIGHT CONTROL
— STORES
SENSOR INFO.
SENSOR DATA MANAGER (FUSION)
KNOWN THREAT DATA
SENSOR INFO.
SITUATION ASSESSMENT
SYSTEM STATUS
EMERGENCY
MISSION PLANNING
TACTICS
PILOT/ VEHICLE INTERFACE
PILOT
"ACTION" SYSTEMS
— FLIGHT CONTROL
— FIRE CONTROL
— ECM
— ENGINE CONTROLS
— WEAPON SYSTEMS
PILOT'S ASSOCIATE

The systems status manager is responsible for monitoring the operational status of all on-board systems, diagnosing potential problems, and projecting the relative "health" of the systems for the remainder of the mission. The status manager will identify any systems that are malfunctioning or which are projected to malfunction. It will then recommend possible actions to compensate for each identified/projected malfunction, such as using a backup or alternative system. For instance, if a navigational instrument fails, the manager may be able to identify the best alternative navigational system to utilize. The system status manager will be capable of taking emergency actions automatically if necessary.

The primary role of the Pilot's Associate is to provide the pilot of a single engine fighter with an automated flight crew that can assume some of the duties of a copilot, weapons operator, navigator and flight engineer. The goal of the program is to develop automated flight crew functions to assist pilot decision-making, rather than to replace pilot function. As a result, the DARPA program is placing heavy emphasis on the development of technologies that will support the highly interactive nature of the pilot/Pilot's Associate interface. Control technologies available to the pilot will include voice inputs, touch-sensitive overlays, programmable switches and HOTAS (hands-on throttle and stick) options. To implement some of these options, the program must of necessity capitalize on research in human speech and natural language understanding generation.

The present focus of the Pilot's Associate program is the development of generic capabilities that are not focused on any particular airframe. The program also has the goal of implementing a capability of a pilot/vehicle interface that will take full advantage of the control, display and automation techniques expected to be part of an advance fighter cockpit projected for around 1995.

10.1.3 Navy Battle Management

As described in Flynn and Senator (1986), the DARPA program in Navy Battle Management includes two components. The first will be an on-shore battle management center at the Fleet Command Center of the Commander-in-Chief of the U.S. Pacific Fleet (CINCPACFLT). The second will be an on-board system aboard the USS Carl Vincent. In both cases, the DARPA program is oriented toward the initial development of prototype systems that can evolve through a series of rapid prototypes into operational systems.

The on-shore system under development is projected as being an integrated battle management decision support system that will have five component application systems: a Force Requirements Expert System (FRESH), a Capabilities Assessment Expert System (CASES), a Campaign Simulation Expert System (CAMPSIM), an Operations Plan Generation Expert System (OPGEN), and a Strategy Generation and Evaluation Expert System (STRATUS). Of these five, only FRESH and CASES appear scheduled for near-term development.

FRESH is oriented toward the monitoring and assessment of current operations. It will monitor the readiness of current CINCPACFLT assets and will assess the impact of any changes in the status of these assets. As significant changes occur, FRESH will evaluate alternative reprioritizations of assets to recommend potential force redirection. It is believed that FRESH has the potential of evolving into a complete resource allocation system for the Pacific Fleet.

CASES will perform specific comparisons of the relative war-fighting capabilities of the U.S. and Soviet Pacific fleets. This comparison will include both a static and a dynamic evaluation. The static evaluation will include an algorithmic comparison of the capabilities of the two opposing fleets based on their respective numbers and capabilities. The dynamic assessment will evaluate mission capabilities on the basis of playing out various scenarios and strategies.

The systems scheduled for installation on-board the USS Carl Vincent will include a Force Level Alerting System (FLASH). FLASH examines correlated reports on various tracks to identify possible threats to the ship, and recommends specific alert conditions based on this information. Initial prototypes of FLASH were developed by Carnegie-Mellon University and the U.S. Naval Ocean Systems Center.

10.1.4 Air Land Battle Management

The focus of the Air Land Battle Management program is to develop a set of cooperative and expandable knowledge-based systems that demonstrate a military planning capability in maneuver control and fire support. Maneuver control involves the planning of movements and actions of ground forces distributed across a battlefield. Fire support involves planning the assignment of distant fires (artillery, missiles, close air support) against enemy targets in support of ground force objectives.

The demonstration system that will be built on this effort will incorporate three cooperating, loosely coupled knowledge-based subsystems that address maneuver control and fire support decisions at the Army Corps level, and maneuver control at the division level. These subsystems are referred to as MOVES(C), FIRES(C) and MOVES(D) respectively. Combined, these three represent three nodes in a conceptual FORCES network. The FORCES network is conceived as being a general architecture that will allow distributed battle management subsystems to be incorporated into a single, integrated battle management system. Consequently, in addition to demonstrating functional planning capabilities in maneuver control and fire support, this program will demonstrate a functional architecture that will gracefully incorporate other knowledge-based battle management subsystems (viz., air defense, combat service support, and intelligence/electronic warfare) at all echelons in the Army (corps, division, brigade, etc.).

Each FORCES node will apply AI techniques to support multiple decision and analysis functions that occur at that node. For instance, for MOVES(C) corps-level maneuver control planning, AI planning techniques may be used to support the generation of alternative courses of actions (COAs); expert systems may be used to support COA evaluation; planning, expert system and natural language processing techniques will be used to support the preparation of detailed orders for lower-echelon identities; expert systems will be used to support situation monitoring, and replanning may once again call on planning techniques.

As with the other strategic computing program efforts, successful development of this system will require advances in certain AI techniques. Specifically, this program calls for the use of parallel processing computers, a conceptual dependency approach to natural language processing, and the development of hierarchical planning techniques that can address the adversarial planning problems characteristic of the military environment.

10.1.5 Research Orientation

The key goal of the Strategic Computing Program is to promote the development of a broad base of machine intelligence technology targeted toward potentially significant enhancements to our national defense systems. The systems discussed above are expected to evolve into a fieldworthy capability. However, the selection of those particular systems was not based solely on their potential operational value, but also largely on the value of the technologies that would be advanced/developed in the process of making these systems a reality. In particular, DARPA efforts will promote significant enhancements in the applicability of expert system and image understanding techniques, while promoting significant technological developments in automated speech recognition and natural language understanding.

To support the very rapid processing requirements of this program's application systems, several key hardware development efforts are also being supported. These efforts are focused in three specific areas: rapid signal processing, rapid symbolic processing, and the development of multiprocessor environments.

To support rapid signal processing, Carnegie-Mellon is working on the development of a programmable systolic array. This is a parallel processor that is extremely fast for computing mathematical functions (fast Fourier transforms, convolution operators) for use in a signal processing environment. A prototype machine called Warp should be able to perform 100 million operations per second. By the time of this publication, Warp is scheduled to be completed.

To support rapid symbolic processing, Texas Instruments is being supported to develop a LISP microprocessor chip that should be able to perform the symbolic functions that now require several hundred microcircuits in today's LISP machines.

More general-purpose, but AI-oriented, parallel processing machines are being developed by corporations such as Bolt, Beranek and Newman (BBN) and Thinking Machines Corporation (TMC). BBN is developing the Butterfly multiprocessor which will be the first of the large-scale multiprocessor architectures to result from this program. The Butterfly will contain 128 processing nodes and will be developed using commercial microprocessors. The Connection Machine by TMC will contain

16,000 microprocessors operating in parallel. Future plans involve the development of a 64,000-node multiprocessor system (eventually up to a million multiprocessors).

10.2 Advanced Digital Radar Imagery Exploitation System

The Advanced Digital Radar Imagery Exploitation System (ADRIES) is a major DARPA-sponsored effort to develop and demonstrate an operational AI capability to exploit ground radar imagery. It represents a single effort that attempts to integrate many of the AI applications noted in Chapter 7. As envisioned, the ADRIES system will provide a set of integrated AI subsystems to (1) utilize image understanding techniques to detect objects (e.g., an armored unit) in a ground radar image, (2) apply AI fusion techniques to correlate the extracted information with other data to update the present picture of the battlefield, (3) further assess the impact of this new information on the overall situation (e.g., enemy intentions), and (4) apply this new assessment to direct the collection of new ground radar imagery.

ADRIES is being developed by Advanced Decision Systems, Inc. along with a number of subcontractors. The ADRIES program was initiated prior to the Strategic Computing Program discussed above. Consequently, more actual development has occurred on this program than on the SCP efforts. However, development is still in a relatively early stage, and the demonstrated capabilities for (2) and (3) above remain somewhat limited.

10.3 Soldier/Machine Interface Program

The Soldier/Machine Interface (SMI) program is a response to a converging pair of trends in the Army, the first technological and the second doctrinal. On the technological side, accelerating advances in compact, powerful and fast computers guarantee that every weapon platform and vehicle will have computational processing capabilities only available in large mainframes a few years ago. Reliable communication capabilities are being projected that suggest the opportunity for real-time data exchange among these microprocessors in a largely interference/interception-free environment.

At the same time that technology is providing information access, doctrinal developments are requiring significant utilization of this information. Doctrinal concepts for the next century postulate a heavily outnumbered U.S. force fighting as a collection of many small, highly mobile, and semiautonomous units coordinated to bring necessary combat power to bear on critical points. This concept implies that the small-unit field commander must be able to utilize the flood of data that technology can provide to determine when and where to move to apply the limited fire power available.

The SMI program is an attempt to develop an on-board capability that will provide the field commander with a single intelligent interface to the various informational inputs. To illustrate the role of an SMI, consider Figure 10-3 which shows the potential sources of information that could constantly bombard a tank platoon commander. The commander could receive information on a variety of different outside channels, any one of which could potentially require his full attention at a given time. In addition, he would need to monitor gauges and indicators displaying the status of on-board sensors and vehicle systems. Some of the gauges would be monitored directly and others indirectly through crew members who would verbally report to him. At the same time, the commander could be receiving constant updates on the overall battlefield situation from tactical fusion centers, and perhaps even data directly downlinked from overhead surveillance. He would have to quickly translate this overall battle situation data into an "over-the-horizon" lookahead to identify and locate a threat before directly perceiving it. Finally, he would be receiving commands or answering queries of higher-echelon commanders while directing and communicating with his subcommanders.

To relieve the field commander of this tremendous information processing burden, an SMI would provide a single sophisticated interface to most incoming data sources (see Figure 10-4), where "sophisticated" means that much of the commander's information processing functions can be offloaded to the SMI. The SMI should be able to filter incoming information for immediate relevance, and present only important data to

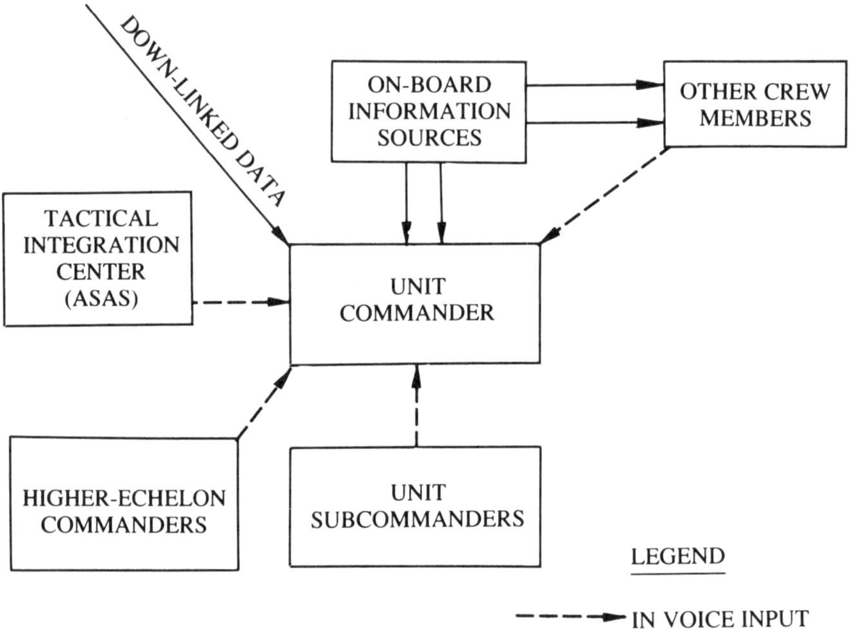

Figure 10-3. Commander's information sources without SMI.

the commander along with suggestions about how to respond to it. Perceptronics, Inc. and BBN are presently developing an SMI prototype under this program.

10.4 Air Force Decision Aids for Command and Control

The Rome Air Development Center (RADC) has sponsored a number of highly-interrelated programs in the development of decision aids for command and control. Combined, these programs represent an attempt to explore all major elements and subelements of a tactical air force (TAF) command and control system and to identify opportunities for incorporating advanced technology decision aids into that environment. While the RADC orientation toward technology is very eclectic ("Use whatever technology works best."), many of the decision aids that have been developed are AI-based.

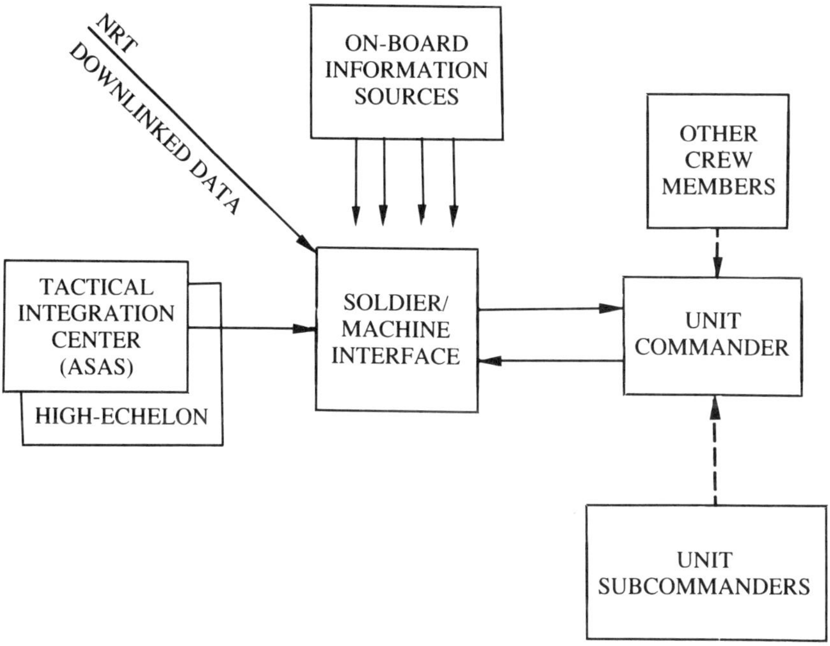

Figure 10-4. Commander's information sources with SMI.

Over the past few years, approximately 15 significant decision aid prototypes have been developed under these programs, most of which were evaluated by operational personnel. Several of these aids are presently moving past the prototype stage into full-scale models with completely functional capabilities.

11
Opportunities, Challenges and Prognoses

As the last four chapters have shown, the Department of Defense is vigorously pursuing many national defense applications of AI. Indeed, the examples given in the last few chapters represent only a small subset of the on-going work in this area. Each case study presented in Chapters 7 through 11 is representative of at least a dozen similar efforts involving similar application domains. In addition, there is considerable work being conducted in many application areas not addressed in this book. Even a casual reading of the Commerce Business Daily, which announces most government procurements, reveals that the applicability of AI is being explored in virtually every area of analysis and decision-making relevant to our national defense. In addition to the few efforts discussed in this book, work is being done in areas such as:

Automated warning of military, political and economic crises

Automated navigation

Computer-aided design
Assembly line robotics
Data management
Financial management
Economic prediction
Logistics planning
Maintenance diagnosis and repair
Battle damage diagnosis and repair
Automated map production
Automated human engineering of systems
Prediction of minefield sites
Intelligent simulation systems
Design of simulation systems

and many others.

Simply stated, AI has emerged as a key technology with widespread applicability and a considerable potential for improving our national defense systems.

11.1 Challenges

While many of these applications look promising, virtually all of the systems addressing them are in a stage of either design or prototype development. Very little of what the scientific community would consider "true AI" has made it beyond the prototype stage, much less into field-worthy operational equipment. As always, making the leap from laboratory to real-world problems will be challenging. This section provides a general discussion of some of the problems AI system developers must overcome before widespread development of practical AI systems in our national defense becomes a reality.

11.1.1 Developing Embedded Expert Systems

As noted in Chapter 1, two types of expert systems can be distinguished: "traditional" expert systems that operate essentially as stand-alone consultation devices, and expert systems that operate as intelligent interfaces. These two types were referred to as *stand-alone expert systems* (SES) and *intelligent interface expert systems* (IES). As depicted in Figure 11-1, the primary difference between an SES and an IES is that an SES assumes a manual input of problem-relevant data while an IES receives its inputs from other automated data sources.

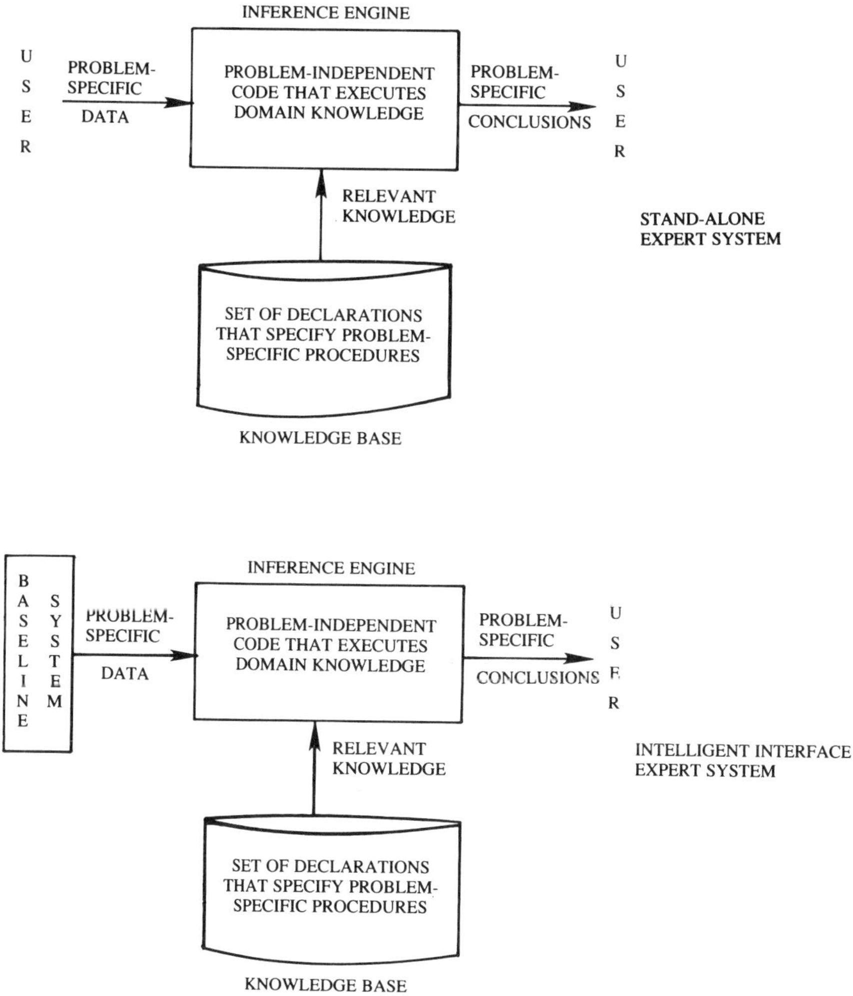

Figure 11-1. The difference between stand-alone and intelligent interface expert systems.

Nearly all national defense applications of expert system technology presented in this book fall into the IES category. In almost every case, the expert system was designed to be embedded within a larger system and served to monitor and process the data generated by that system. In addition, much of the work in image understanding assumes an architecture similar to IES architecture. The military emphasis on embedded expert systems contrasts sharply with academic experience, where the traditional focus has been on stand-alone systems.

It is argued here that implementation of an embedded expert system or IES is a very different problem than that of developing a stand-alone expert system. Furthermore, it is suggested that these differences suggest an approach to IES implementation that may be somewhat different from that which is appropriate for SES implementation. Specifically, there are three categories of differences: software engineering, knowledge engineering, and user interface development.

Software Engineering

An IES is usually designed to support decision problems that are significantly time-constrained. This is largely an inherent aspect of IES application domains, since the need for the automatic computation of primitive values is generally due to the lack of time available to enter data manually. The decision domains appropriate for SES applications, such as medical diagnosis, are often characterized by a relatively time-relaxed environment where the decision-maker has some opportunity to ponder the problem. The AASW RTAS described in Chapter 8 is a typical example.

The need for rapid advice generation in IES systems poses two problems. First, there is the requirement that the inference engine must be capable of processing a large number of rules quickly. Second, the IES must have embedded within it not only the inference engine, but also a mechanism for computing and managing primitive values. Keeping a current record of such values can be a computationally extensive process in cases where the output of the baseline data source(s) and the primitive value input requirements of the IES do not coincide. Unfortunately, as discussed in the following section, this is generally the case.

Currently, two approaches exist for resolving this problem. First is the development of very fast LISP processing environments. LISP is the most commonly used AI development language in the U.S. If AI system developers are going to remain with LISP, then much faster LISP processing capabilities will be required for real-time applications. Another approach is to use a more "standard" and structured language for development, such as Pascal, C or Ada. In such languages, the developer could, for instance, write a compiler that translates knowledge declared as production rules into a set of source code functions that are further compiled into object code at run-time. While this approach loses much of the flexibility that an interpretive LISP environment provides, it does provide for the rapid execution of rules and is easily expanded to include functions for computing primitive values. It also has the advantage of allowing expert system development in the same language that is becoming the standard for other DoD applications, namely Ada.

Knowledge Engineering

Knowledge engineering is the process of defining and encoding the contents of a knowledge base. The traditional approach to knowledge engineering is to find a person who is an established expert at solving the problems the expert system will address, and identifying the heuristics the human expert uses to solve those problems. In effect, the expert system mimics the procedures of a human expert. Unfortunately, for many military decision problems, experts of this type may not be available for two reasons. First, many IES applications are specifically designed to support functions that cannot be performed manually. For instance, an IES application may involve the rapid processing and interpretation of large volumes of data generated by various sensor systems. Such applications require processing speeds people cannot handle. This is not a comment on the limitations of human reasoning, but rather reflects the careful selection of problems to which machines may be uniquely suited. A second reason is simply that the IES application may be part of the development of a new system, and so human experts with experience in using the baseline system do not exist. Note that the IES applications embedded within the Pilot's Associate program have both of these characteristics.

With no experts to mimic, the contents of a rule base cannot be validated with expert judgments. Evaluation procedures that compare human and machine reasoning are thus no longer valid. Consequently, objective procedures for measuring the quality of knowledge base elements are necessary for both knowledge engineering and knowledge base validation. Few objective procedures exist today for measuring the quality of a knowledge base.

Human Engineering

Despite the fact that IESs fall into the category of "intelligent" interfaces, they pose some interesting human engineering problems that emerge from the need of an IES to justify its advice. It is not sufficient for an IES to simply recommend actions; it must also be able to explain to a user reasons for its recommendations. With expert systems that fall into the SES class, this is traditionally done by displaying a trace of the rules. With many intelligent interface applications of expert system technology, however, several factors combine to make this approach both inadequate and infeasible.

One factor is the generally diverse nature of the user community of an IES. With an SES, the requirement that data be manually entered is often sufficient to guarantee a moderate level of problem domain expertise in a user. With an IES, automated data entry occurs, resulting in an essentially turnkey system and the possibility that novice users without significant domain expertise will utilize the system. Consequently, IES justifications of advice cannot presume that the user is familiar with the heuristics embedded in the knowledge base. A second factor is that since good IES knowledge engineering is not based on mimicking human expertise, it is likely that the problem perspective embedded in the IES reasoning process will be somewhat different than the perspective of even domain-expert users. The advice justification component must be able to account for this difference. The third and perhaps most severe factor is the constraint imposed on the time available to an IES to justify its advice. For real-time applications, IES justifications must be understandable "at a glance". This difficulty is compounded by the fact that the justification must trace both the rules that lead to a result *and* the relevant primitives that were computed. This is significantly different from an SES where it can be presumed that the user is a priori familiar with the primitive values by the fact that the user entered them.

Overall, the design of advice justification in an SES appears much easier than with an IES. With an SES, the only thing to display is a rule trace; the user entered everything else. With an IES, advice justification must include both rules and data, cannot presume user understanding of rules when they are presented, and must be done in a few seconds. Developing user interfaces with this capability remains a difficult challenge.

11.1.2 Extensive Knowledge Development

AI systems are knowledge-based. In every AI system there exists a file that provides a representation of the elements of a problem domain that provides the AI system with the base of knowledge necessary to make decisions and act within that domain. Early research experience in building AI systems generally involved proof-of-principle demonstrations with test problems of very limited scope. As AI systems move out of these test domains into practical applications, it is often discovered by the system developers that the quantity of knowledge required by the system is one or more orders of magnitude greater than expected. In rule-based systems, for instance, some estimate that many "large scale" problem domains (such as battle management) will require knowledge bases with 10,000 or more rules. Most systems that have been developed to date address limited "small scale" problem domains, and have rarely exceeded 1,000 rules.

Knowledge engineering is the process of extracting and encoding knowledge. As discussed in Chapter 1, this process is often labor-intensive and very time-consuming. Some estimates range as high as two man-weeks of effort for each rule in the knowledge base. If this estimate is true, then a 10,000-rule system would require in excess of 350 man-years to develop. Even at the rate of one rule per workday, knowledge engineering would still require in excess of 35 man-years of effort.

If AI systems are to be scaled-up to many of the practical applications the developers of these systems hope to address, then the process of knowledge engineering must become a great deal more efficient and streamlined than is true of present techniques. This represents a major challenge to the practical application of AI technology.

11.1.3 Merge AI and Traditional System Development Methodologies

Within AI, particularly with expert systems, the accepted development methodology is that of rapid prototyping and frequent iteration. Since the functionality of an expert system may often change with modifications to the knowledge base, it is often impossible to completely design the system before beginning development. Traditional structured software development techniques, however, generally specify a well-defined sequence of stages in software implementation. These stages include (1) a functional requirements analysis, where the functionality of the system is precisely defined, (2) a structured design, where all modules and algorithms in the system are carefully described and documented, and (3) development, where all modules are actually coded, usually in a top-down fashion.

While structured development techniques have proven generally unacceptable for AI system development, it remains a fact that most of the AI systems proposed for national defense applications will be embedded within much larger software/hardware systems, and that the development of these larger systems will be done using structured design and development methodologies. If the AI component is essential to the operation of the larger system, then it is unlikely that the overall system developer will accept imprecise functional descriptions of the AI components of that system.

Resolving this conflict between AI and more traditional development approaches is necessary before the development of essential AI subsystems become generally accepted.

11.1.4 Moving Beyond the Prototype Stage

Over the past 20 years, literally hundreds of prototype decision aids have been developed that successfully demonstrated the utility of AI and other advanced analytic technologies (e.g., decision analysis and operations research) for supporting battle management decision-making and intelligence analysis. Often, the utility of the aid is demonstrated empirically by comparing aided and unaided decision-making performance. Despite this, very few decision aids have moved beyond the prototype stage into operational environments. A number of explanations have been given to account for this—difficulties/slowness in the government procurement process, lack of user acceptance, mismatches between the proposed aiding system and the requirements of the decision-making organization, poor marketing of the aiding concept, etc. Whether any or all of these explanations are valid, it is clear that the lack of transferability of operational environments remains an on-going concern. If AI systems are ever to have widespread acceptance in the operational national defense community, this is one issue that must clearly be resolved.

11.2 Prognoses for the Future

In this book, we have explored the applicability of AI technology to our national defense. Basic tools and techniques of AI were presented, along with descriptions of a variety of prototype application systems and DoD initiatives. Opportunities to apply AI technology and challenges to the successful exploitation of those opportunities were discussed. As we have discovered, both the opportunities and challenges are quite numerous. In an effort to be objective, I have avoided making assessments about whether any individual systems or research programs are likely to lead to the development of operational systems. As my final comments, I would therefore like to present my own view on both near-term and long-term prospects for AI in our national defense.

My first prediction is that most of the national defense applications of AI presently being pursued will not succeed in the near-term development of operationally useful systems, despite the fact that many of these programs have the specific objective of developing operationally functional systems in the near future. Many if not most of these programs are *technology-push* programs that specify a priori that AI techniques will be used. Most CBD announcements for these programs begin with phrases such as "Apply AI to . . . ", "Build an expert system to . . . ", "Develop an AI system that is capable of . . . ", as opposed to more requirements-oriented phrases such as, "Find a solution to the problem of . . . ", "Develop a technique that will . . . ", etc. While technology-push efforts are very important in their own right, it must be recognized that the goal of such programs is to *attempt* to apply the specified technology to the selected prospect, and not to solve the problem per se. Often, it is not initially clear whether the application of the technology is feasible, and the goal of the development effort may well be to determine feasibility. Furthermore, potential solutions and technical concepts that fall outside the scope of the specified technology are also outside the scope of the effort. As a result, the fact that AI applications are being widely pursued should not lead one to conclude that strong evidence exists that these techniques will succeed. In fact, quite the opposite is true—there is no substantial evidence that AI techniques are sufficient for solving the problems that are being addressed. The "challenges" to AI that were discussed in this and previous chapters were identified as challenges largely because of past failures to apply AI techniques. At the same time, few if any AI system development efforts have been so successful as to demonstrate that these difficulties can be overcome. For these reasons, I believe it only reasonable not to expect too much from these on-going programs.

The fact that most of these efforts will eventually fall short as operational systems will also lead to a significant decrease in the amount of funding available for AI system development. Many members of the AI community believe that an "AI winter" is inevitable. Indeed, if it were not for the challenge of the Japanese Fifth Generation Computing Program, some believe this AI winter would have already occurred.

On the other hand, some of these programs will be successful. The simple fact that there exist numerous programs in many different problem domains leads one to expect some successes. In my opinion, there is no evidence of "fatal flaws" in AI theories and approaches that could lead one to believe otherwise. With success as well as failure, we will learn more about what it takes to build operational systems based on AI technology. From this learning experience, a methodology will begin to emerge that will describe the steps necessary for developing operational systems. At the same time, government sponsors will become more knowledgeable and selective about the programs to which AI techniques are to be applied. The so-called "AI winter" is only a matter of funding. While the total number of AI efforts may decrease dramatically, the proportion of successful AI efforts will increase just as dramatically.

In short, for the next decade or so, I envision a period of many failures with some scattered successes along with a significant decrease in research and development funding.

The long-term trend is quite a different story. There is no doubt that eventually there will exist intelligent machines capable of performing the analytic and decision-making functions discussed in this book, and probably a great deal more. By that time, however, today's apparent romance with AI will have dissipated. The tools and techniques of AI will be perceived as no more special than those of, say, data base management systems. Artificial intelligence in machines will no longer be something extraordinary, but simply part of everyday life.

References

Alvey Committee, "A Programme for Advanced Information Technology: The Report of the Alvey Committee," HMSO, 1982.

Andriole, S., Hopple, G., Thompson, J, and Bloch, H. "An Intelligent Aid for Tactical Planning," *IEEE Transactions of Systems, Man, and Cybernetics*, November/December 1986.

Barr, A. and Feigenbaum, E.A. (eds.). *The Handbook of Artificial Intelligence: Vol. 1*. William Kaufmann, Inc.: Los Altos, CA, 1981.

Barr, A. and Feigenbaum, E.A. (eds.). *The Handbook of Artificial Intelligence: Vol. 2*. William Kaufmann, Inc.: Los Altos, CA, 1982.

Barstow, D. "A Perspective on Automatic Programming," *AI Magazine*, 1984, Vol. 5, No. 1, pp. 5-27.

Barth, S.W., Coyle, H. and Sobik, D. "User Guide for the Duplex Army Radio/Radar Targeting Aid (DART)," PAR Technology Corporation Report, 1984.

Barth, S.W. and Lehner, P.E. "Artificial Intelligence on Microcomputers," in

Microcomputer Decision Support Systems, ed. S. Andriole, QED Information Systems: Cambridge, MA, 1986.

Berliner, H.J. "Some Necessary Conditions for a Master Chess Program," *Proceedings of the Third International Joint Conference on Artificial Intelligence*, Stanford University, Stanford, CA, 1973.

Bledsoe, W.W. "Splitting and Reduction Heuristics in Automatic Theorem Proving," *Artificial Intelligence*, 1971, Vol. 2, No. 1, pp. 55-77.

Bledsoe, W.W. and Bruell, P. "A Man-Machine Theorem Proving System," *Artificial Intelligence*, 1974, Vol. 5, No. 1, 51-72.

Bramer, M. and Bramer, D. *The Fifth Generation: An Annotated Bibliography*, Addison-Wesley Publishing Company: Reading, MA, 1984.

Brooks, R.A. "Symbolic Reasoning Among 3-d Objects and 2-d Models," *Artificial Intelligence*, 1981, Vol. 16, pp. 285-348.

Carbonell, J.R. "AI in CAI: An Artificial Intelligence Approach to Computer-Assisted Instruction," *IEEE Transactions on Man-Machine Systems*, 1970, Vol. 11, pp. 190-202.

Carr, B. and Goldstein, I., "Overlays: A Theory of Modeling for Computer Aided Instruction," AI Memo 406, AI Laboratory, Massachusetts Institute of Technology, 1977.

Christiansen, A.D. "The History of Planning Methodology: An Annotated Bibliography," *SIGART Newsletter*, October, 1985.

Clancey, W.J. "The Epistemology of a Rule-Based Expert System: A Framework for Explanation," *Artificial Intelligence*, 1984.

Clark, K.L. and McCabe, F.G. "Prolog: A Language for Implementing Expert Systems," in *Machine Intelligence 10*, eds. Hayes, J.E., Michie, D. and Pao, Y-H., Halsted Press: John Wiley and Sons, New York, 1982.

Clocksin, W.F. and Mellish, C.S. *Programming in Prolog*, Springer-Verlag, New York, 1981.

Cohen, P.R. and Feigenbaum, E.A. (eds.). *The Handbook of Artificial Intelligence: Vol. 3*. William Kaufmann, Inc.: Los Altos, CA, 1982.

D'Amore, R.J. and Mah, C.P. "One Time Complete Indexing of Text: Theory and Practice," *Conference on Research and Development in Information Retrieval*, Cambridge, England, June, 1984.

Doyle, J. "A Truth Maintenance System," *Artificial Intelligence*, Vol. 1, No. 3, 1979.

Duda, R.O., Hart, P.E. and Gashing, J. "Model Design in the PROSPECTOR Consultant System for Mineral Exploration," in *Expert Systems in the Microelectronic Age*, ed. Michie, D., Edinburgh University Press, Edinburgh, 1979.

Dyer, N.L, Frantz, F.K. and Livingston, M. "AELASS: A Knowledge-Based System for Multisensor Surveillance Data Exploitation," *Proceedings of the Conference on Artificial Intelligence*, Oakland University, Rochester, MI, April 26-27, 1983.

Engleman, C., Scarl, E.A. and Berg, C.H. "KNOBS: An Experimental Knowledge-Based Tactical Air Mission Planning System and A Rule-Based Aircraft Identification Simulation Facility," *Proceedings of the Sixth Annual Joint Conference on Artificial Intelligence*, 1979, pp. 247-249.

Feigenbaum, E.A. and McCorduch, P. *The 5th Generation*, Addison-Wesley: Reading, MA, 1983.

Fields, C. "The Strategic Computing Program: An Overview," *SIGNAL*, August, 1984.

Fikes, R.E., Hart, P.E., and Nilsson, N.J. "Learning and Executing Generalized Robot Plans," *Artificial Intelligence*, 1972, Vol. 3, No. 4, pp. 251-288.

Fletcher, J.D. "Intelligent Instructional Systems in Training," in *Applications in Artificial Intelligence*, ed. S. Andriole, Petrocelli Books, Inc.: Princeton, NJ, 1985.

Fletcher, J.D. and Psotka, J. "Intelligent Training Systems for Maintenance," *SIGNAL*, 1986, Vol. 40, No. 10, pp. 89-95.

Flynn, J.P. and Senator, T.E. "DARPA Navy Battle Management Applications," *SIGNAL*, Vol. 40, No. 10, pp. 59-70.

Friedland, P.E. "Knowledge-based Experiment Design in Molecular Genetics," Doctoral dissertation, Computer Science Department, Stanford University, Stanford, CA, 1979.

Gevarter, W.B. "Expert Systems: Limited But Powerful," in *Applications in Artificial Intelligence*, ed. S. Andriole, Petrocelli Books, Inc.: Princeton, NJ, 1985.

Granger, R.H., Meyers, A., Taylor, G.B. and Yoshi, R. "NOMAD: A Naval Message Understanding System," *Proceedings of the Conference on Artificial Intelligence*, Oakland University, Rochester, MI, April 26-27, 1983.

Hayes-Roth, B. and Hayes-Roth, F. "A Cognitive Model of Planning," *Cognitive Science*, Vol. 3., 1979, pp. 275-310.

Hayes-Roth, F., Waterman, D.A. and Lenat, D.B. (eds.). *Building Expert Systems*, Addison Wesley: Reading, MA, 1983.

Harris, L.R. "User Oriented Data Base Query with Robot Natural Language Query System," *Proceedings of the Third International Conference on Very Large Data Bases*, Tokyo, October, 1977.

Harris, M.D. *Introduction to Natural Language Processing*, Reston Publishing Company, Inc.: Reston, VA, 1985.

Hendrix, G. and Sacerdoti, E. "Natural Language Processing: The Field in Perspective," *BYTE*, September, 1981, pp. 304-352.

Hendrix, G., Sacerdoti, E., Sagalowicz, D., and Slocum, J. "Developing a Natural Language Interface to Complex Data," *ACM Transactions on Database Systems*, 1978, Vol. 3, pp. 105-107.

JIPDEC. "Preliminary Report on Study and Research on Fifth Generation Computers 1979-1980," Japan Information Processing Development Center, 1981.

Kaplan, S. "Cooperative Responses from a Portable Natural Language Data Base Query System," Ph.D. dissertation, University of Pennsylvania, 1979.

Kelly, C. "Autonomous Robot Vehicle Program," *SIGNAL*, August, 1984, pp. 104-106.

Lehner, P.E. "Planning Under Adversity: A Computational Model of Strategic Planning in the Game of Go," Doctoral dissertation, Department of Psychology, University of Michigan, Ann Arbor, MI, 1981.

Lehner, P.E. "Strategic Planning in Go," in *Computer Game Playing: Theory and Practice*, ed. M. Bramer, Ellis Harwood Ltd: England, 1983.

Lehner, P.E., McIntyre, J.R., Adelman, L.A., Gates, K., Luster, P.W., Probus, M. and Donnel, M.L. "Combining Decision Analysis and Artificial Intelligence Techniques: An Intelligent Aid for Estimating Enemy Courses of Action," PAR Report No. 84-131, March, 1984, PAR Technology Corporation.

Lehner, P.E., Probus, M.R. and Donnel, M.L. "Building Decision Aids: Exploiting the Synergy Between Decision Analysis and Artificial Intelligence," *IEEE Transactions of Systems, Man, and Cybernetics*, SMC-15(14), July/August, 1985.

Lehner, P.E. and Barth, P.E. "Expert Systems on Microcomputers," *Expert Systems: The International Journal of Knowledge Engineering*, Vol. 2, No. 4, October, 1985, pp. 198-208.

Lehner, P.E. "Adversarial Planning Procedures with Provable Properties," in preparation, 1988.

Lehner, P.E. and Zirk, D.A. "Cognitive Factors in User/Expert System Interaction," *Human Factors*, in press.

Lemmer, J.F. and Mitchell, B.T. "Interactive Aids for the Development of Computer Vision Rule Bases," PAR Technology Report, 1984.

Mah, C.P. "Advanced Query Facility," PAR Report No. 80-53, 1980, PAR Technology Corporation.

Martin, W.A. and Fateman, R.J. "The MACSYMA System." *Proceedings of the 2nd Symposium on Symbolic and Algebraic Manipulation*, Los Angeles, CA, 1971, pp. 23-25.

Martin, J. *Application Development Without Programmers*, Prentice-Hall, Inc.: Englewood Cliffs, NJ, 1982.

McWilliams, E.D. and Ricard, G.L. "Portable, Intelligent Simulation for ASW Training," in *Computer-Based Instruction for Military Environments*, North Holland Press: Amsterdam, to be published.

Michalski, R.S., Carbonell, J.G., and Mitchell, P.M. *Machine Learning: An Artificial Intelligence Approach*, Tioga Publishing Co.: Palo Alto, CA, 1983.

Michalski, R.S., Carbonell, J.G., and Mitchell, T.M. *Machine Learning: An Artificial Intelligence Approach: Vol. II*, William Kaufmann Publishers, Inc.: Los Altos, CA, 1986.

Pecora, V.J. "EXPRS: A Prototype Expert System Using Prolog for Data Fusion," *AI Magazine*, Vol. 5, No. 2, 1984, pp. 37-41.

Pitrat, J. "A Chess Program Which Uses Plans," *Artificial Intelligence*, Vol. 8, 1977, pp. 275-321.

Reitman, W. and Wilcox, B. "Modeling Tactical Analysis and Problem Solving in Go," *Proceedings of the Tenth Annual Pittsburgh Conference on Modeling and Simulation*, Pittsburgh, PA, 1979, pp. 2133-2148.

Rich, E. *Artificial Intelligence*, McGraw-Hill: New York, 1985.

Sacerdoti, E.D. "The Structure of Plans and Behavior," Doctoral dissertation, Computer Science Department, Stanford University, Stanford, CA, August, 1975.

Sacerdoti, E.D. and Sagalowicz, D. "A LADDER User's Guide," SRI International, Technical Note 163R, March, 1980.

Schank, R.C. and Abelson, R.P. *Scripts, Plans, Goals, and Understanding*. Lawrence Erlbaum: Hillsdale, NJ, 1977.

Shortliffe, E.H. *Computer-Based Medical Consultations: MYCIN*, American Elsevier: New York, 1976.

Slagle, J.R. and Hamburger, H. "An Expert System for a Resource Allocation Problem," *Communication of the ACM*, 1985, Vol. 28, No. 9, pp. 994-1004.

Stevens, A.L., Collins, A. and Goldin, S. "Diagramming Students' Misconceptions in Causal Models," BBN report no. 3786, Bolt, Beranek and Neumann, Inc., 1978.

Sussman, G.J. *A Computer Model of Skill Acquisition*, American Elsevier: New York, 1975.

Waltz, D. "Generating Semantic Descriptions from Drawings of Scenes with Shadows," in *The Psychology of Computer Vision*, ed. P. Winston, McGraw-Hill: New York, 1975.

Waltz, D. "Initial Language Access to a Large Data Base: An Engineering Approach," *Proceedings of the 4th International Joint Conference on Artificial Intelligence*, Tibilisi, Georgia, USSR, pp. 868-872.

Wilensky, R. *Planning and Understanding: A Computational Approach to Human Reasoning*, Addison-Wesley: Reading, MA, 1983.

Wilkins, D. "Using Patterns and Plans to Solve Problems and Control Search," Stanford Artificial Intelligence Laboratory Memo AIM-329, Stanford University, 1979.

Wilkins, D.E. "Domain Independent Planning: Representation and Plan Generation," *Artificial Intelligence*, Vol. 22, 1984, pp. 269-301.

Woods, W.A. "Transition Network Grammars for Natural Language Analysis," *Communication of the ACM*, 1970, Vol. 13.

Young, P.R. and Lehner, P.E. "Applications of a Theory of Automated Adversarial Planning to Command and Control," *IEEE Transactions on Systems, Man, and Cybernetics*, November/December, 1986.

Index

Index

D

E

F

G

H

I

J

K

N

O

P

U

V

W

X